Biotechnology
Intelligence
Unit

Recombinant Cell Surface Receptors: Focal Point for Therapeutic Intervention

Michael J. Browne

SmithKline Beecham Pharmaceuticals
Harlow, Essex, United Kingdom

Academic Press

R.G. Landes Company
Austin

Biotechnology Intelligence Unit

RECOMBINANT CELL SURFACE RECEPTORS: FOCAL POINT FOR THERAPEUTIC INTERVENTION

R.G. LANDES COMPANY
Austin, Texas, U.S.A.

This book is printed on acid-free paper.

Please address all inquiries to the Publisher:
R.G. Landes Company
810 S. Church Street, Georgetown, Texas, U.S.A. 78626
Phone: 512/ 863 7762; FAX: 512/ 863 0081

Academic Press, Inc.
525 B Street, Suite 1900, San Diego, California, U.S.A. 92101-4495

United Kingdom Edition published by Academic Press Limited
24-28 Oval Road, London NW1 7DX, United Kingdom

Library of Congress Catalog Number: 615'.19—dc20
International Standard Book Number (ISBN): 0-12-137765-2

Printed in the United States of America

Library of Congress Cataloging-in-Publication Data

Browne, M. J. (Michael J.)
Recombinant cells surface receptors: focal point for therapeutic intervention/ Michael J. Browne
p. cm. — (Biotechnical intelligence unit)
Includes bibliographical references and index.
ISBN 1-57059-410-4 (alk. paper)
1. Drug receptors. 2. Recombinant proteins. 3. Receptor-ligand complexes. I. Title. II. Series.
RM301.41.B76 1996
615'.19—dc20 95-23860
CIP

Publisher's Note

R.G. Landes Company publishes six book series: *Medical Intelligence Unit, Molecular Biology Intelligence Unit, Neuroscience Intelligence Unit, Tissue Engineering Intelligence Unit, Biotechnology Intelligence Unit* and *Environmental Intelligence Unit.* The authors of our books are acknowledged leaders in their fields and the topics are unique. Almost without exception, no other similar books exist on these topics.

Our goal is to publish books in important and rapidly changing areas of bioscience for sophisticated researchers and clinicians. To achieve this goal, we have accelerated our publishing program to conform to the fast pace in which information grows in bioscience. Most of our books are published within 90 to 120 days of receipt of the manuscript. We would like to thank our readers for their continuing interest and welcome any comments or suggestions they may have for future books.

Shyamali Ghosh
Publications Director
R.G. Landes Company

CONTENTS

EDITOR

Michael J. Browne, Ph.D.
SmithKline Beecham Pharmaceuticals
New Frontiers Science Park
Third Avenue
Harlow, Essex, United Kingdom
Chapter 1

CONTRIBUTORS

Allan Blake, Ph.D.
Department of Pharmacology
University of Pennsylvania School of Medicine
Philadelphia, Pennsylvania, U.S.A.
Chapter 3

Joël Bockaert
CNRS UPR 9023
Mécanismes Moléculaires des Communications Cellulaires
CCPIE
Montpellier Cedex, France
Chapter 5

Olivier Civelli, Ph.D.
Department of Pharmacology
College of Medicine
University of California, Irvine
Irvine, California, U.S.A.
Chapter 2

Regis Grailhe
Center for Neurobiology and Behavior
Columbia University
New York, New York, U.S.A.
Chapter 6

Mark J.S. Heath, M.D.
Center for Neurobiology and Behavior
Columbia University
New York, New York, U.S.A.
Chapter 6

René Hen, Ph.D.
Center for Neurobiology and Behavior
Columbia University
New York, New York, U.S.A.
Chapter 6

Anthony G. Hope, Ph.D.
The Neurosciences Institute
Department of Pharmacology and Clinical Pharmacology
Ninewells Hospital and Medical School
The University of Dundee
Dundee, United Kingdom
Chapter 7

Jeremy J. Lambert, Ph.D.
The Neurosciences Institute
Department of Pharmacology
and Clinical Pharmacology
Ninewells Hospital and Medical
School
The University of Dundee
Dundee, United Kingdom
Chapter 7

Jose J. Lucas, Ph.D.
Center for Neurobiology
and Behavior
Columbia University
New York, New York, U.S.A.
Chapter 6

Ruth M. McKernan, Ph.D.
Neuroscience Research Centre
Merck Sharp and Dohme Research
Laboratories
Terlings Park
Harlow, Essex, United Kingdom
Chapter 8

Frederick J. Monsma Jr., Ph.D.
Pharma Research, Preclinical
Neurosciences
F. Hoffman-La Roche
Basel, Switzerland
Chapter 2

Danuta Mossakowska, Ph.D
SmithKline Beecham Pharmaceuticals
New Frontiers Science Park
Harlow, Essex, United Kingdom
Chapter 10

Masataka Nakamura, Ph.D.
Human Gene Sciences Center
Tokyo Medical and Dental
University
Tokyo, Japan
Chapter 9

Ronald S. Oosting, Ph.D.
Center for Neurobiology &
Behaviour
Columbia University
New York New York, U.S.A.
Chapter 6

John A. Peters, Ph.D.
The Neurosciences Institute
Department of Pharmacology
and Clinical Pharmacology
Ninewells Hospital and Medical
School
The University of Dundee
Dundee, United Kingdom
Chapter 7

Jean-Philippe Pin
CNRS UPR 9023
Mécanismes Moléculaires des
Communications Cellulaires
CCPIE
Montpellier Cedex, France
Chapter 5

Terry Reisine, Ph.D.
Department of Pharmacology
University of Pennsylvania School
of Medicine
Philadelphia, Pennsylvania, U.S.A.
Chapter 3

Richard A.G. Smith, Ph.D.
SmithKline Beecham
Pharmaceuticals
New Frontiers Science Park
Harlow, Essex, United Kingdom
Chapter 10

A.D. Strosberg, Dr. Sci.
Institut Cochin de Génétique
Moléculaire
Laboratoire d'Immuno-
Pharmacologie Moléculaire
CNRS UPR 0415-22
Paris, France
Chapter 4

Kazuo Sugamura, M.D., Ph.D.
Department of Microbiology
Tohoku University School of
Medicine
Sendai, Japan
Chapter 9

Lesley Sutherland, Ph.D.
The Neurosciences Institute
Department of Pharmacology
and Clinical Pharmacology
Ninewells Hospital and Medical
School
The University of Dundee
Dundee, United Kingdom
Chapter 7

Melanie Tallent, Ph.D.
Department of Pharmacology
University of Pennsylvania School
of Medicine
Philadelphia, Pennsylvania, U.S.A.
Chapter 3

Paul J. Whiting, Ph.D.
Neuroscience Research Centre
Merck Sharp and Dohme Research
Laboratories
Terlings Park
Harlow, Essex, United Kingdom
Chapter 8

CHAPTER 1

Introduction: Recombinant Cell Surface Receptors—Focal Point for Therapeutic Intervention

Michael J. Browne

This book encapsulates the rapid progress now being made in the identification and use of recombinant cell surface receptors. The approach has been to identify a number of receptor systems of well recognized, high medical value (either realized or potential) and invite key research workers in both academia and the pharmaceutical industry to highlight significant progress both at a basic research level and in terms of clinical benefit.

Receptors provide a key point for intervention in, and understanding of, numerous, diverse, important pathological states ranging from psychiatric disorders such as schizophrenia, neurodegenerative disorders such as Parkinsonism, dysfunction of receptors involved in the immune response leading to allergy or impaired immunity, to cardiovascular dysfunction. The key role of receptors in the prospects for new, improved therapeutic intervention strategies justifies the increasing intensity of research in academia and the pharmaceutical industry.

The structure of mammalian cell surface receptors varies widely, as does the number of subunits: one in the case of the seven-transmembrane domain G-protein linked class (7TM), probably a heteropentamer in the case of the GABA receptor. The mechanism of interaction with the cognate intracellular signaling system is also quite varied: for example, coupling to intracellular G proteins or gating of an ion channel.

The main unifying theme, however, is that receptor-based research has a distinguished track-record of successful follow-through to clinical

Recombinant Cell Surface Receptors: Focal Point for Therapeutic Intervention, edited by Michael J. Browne.

application. In the case of the opiate receptors, it is clear that analgesic agents acting on these receptors have been in use for several millennia. More recent rational receptor-based strategies have spawned numerous productive avenues of research, leading, for example, to H_2 antagonists used in treating stomach ulcers or β1–adrenoceptor blockers used in angina and hypertension and β2–adrenoceptor agonists used in treating asthma.

The advent of recombinant DNA technology provides the means to build further on these initial successes. Significantly, recombinant technology offers: access to defined preparations of 'homogeneous' receptor (perhaps not previously defined by pharmacology), particularly human, types or subtypes; the prospect of studying the composition and functionality of multisubunit heteromeric receptors; a means of examining the interaction with signaling systems in different host cell backgrounds, the tools for highly efficient drug screening strategies and the prospect of high level expression to underpin structural studies. The latter, in principle, provides the means for improved rational de novo drug design.

In some circumstances recombinant receptors themselves may be employed, sometimes in modified form, as therapeutic agents—in effect soluble 'decoy' receptors to 'mop up' circulating agonists.

In ethical terms, the increasing availability of recombinant receptors provides a means of reducing the reliance on animal sources of tissue for experiments in vitro, with the further prospect that more focused studies in vitro may reduce the need for experiments in vivo. Additionally inter-species comparisons are also more readily made, helping the transition between preclinical studies in vivo to trials in man.

In the last decade recombinant DNA technologies have fueled a race to isolate novel receptors. This has been particularly productive in the area of the serotonin and dopamine receptors. In many instances researchers specifically sought out these receptors based on a known pharmacological activity or perceived anomalies in pharmacology or by searching for human homologs of known animal receptors. More recently the launch of large-scale, broadly based cDNA cloning efforts, signally exemplified by the work of Adams et al,[1] have revealed numerous novel genes. By strategies such as these, our own laboratories already have access to over 40 novel 7TM receptors.[2]

By the very nature of this recent 'gene first' strategy the endogenous ligand and hence the pharmacology of the novel receptor will initially, more often than not, be unknown. Having the receptor *first* provides a new and exciting starting point to define entirely new endogenous ligands[3] and new physiological and pharmacological processes. Even when the 'Human Genome Project' has revealed the full coding potential of the human genome, scheduled variously between the years 2003 and 2005, much will still remain to be elucidated in terms of the biology of the novel receptor systems, their structures and interaction with ligands and signal transduction systems and the eventual realization of the potential medical benefit.

In addition to the potential to develop therapeutic agents, efforts are also in place to focus diagnostic approaches much more specifically. Indeed there appears to be reason for optimism that even complex multifactorial disorders may yield to focused studies on well defined patient populations–such as the recent demonstration of an association between schizophrenia and a variant of the human 5HT2a receptor.[4] Teaming of novel diagnostic tests with novel (or even existing) therapeutic agents could provide for more effective targeting of and economic treatment of numerous human ailments.

Receptor-based medicinal research has served us well to date and promises to continue to do so well into the future.

REFERENCES

1. Adams M, Kerlavage A R, Fleischmann R D et al. Genome Directory. Nature 1995; 377(Supp):3-174.
2. Edgington SM. Editorial. Biotechnology 1995; 13:1152-1153.
3. Meunier J-C, Mollereau C, Toll L et al. Isolation and structure of the endogenous agonist of opioid receptor-like ORL1 receptor. Nature 1995; 377:532-535.
4. Williams J, Spurlock G, Laurent C et al. Association between schizophrenia and T1-2C polymorphism of the 5-hydroxytryptamine type 2a-receptor gene. Lancet 1996; 437(9011):1294-6.

CHAPTER 2

Dopamine Receptors

Frederick J. Monsma, Jr. and Olivier Civelli

2.1 INTRODUCTION

Interest in the dopaminergic neurotransmitter system dates back to the late 1950s when Carlsson and colleagues demonstrated the presence of dopamine in the brain and showed that the depletion of catecholamines by reserpine could be antagonized by the precursor of dopamine, 3,4-dihydroxyphenylalanine (L-DOPA).[1,2] The significance of this finding was fully realized when it was demonstrated that dopamine was depleted in the brains of Parkinson's patients,[3] and that administration of L-DOPA could reverse Parkinsonian symptoms dramatically.[4,5] Further interest in the dopamine system was generated following its implication in psychiatric disease, initially suggested in 1963-4 by the observation that the highly effective antipsychotic drugs chlorpromazine and haloperidol affected monoamine metabolism, but not monoamine levels.[6] This effect was due presumably to the blockade of monoaminergic receptors leading to compensatory activation of monoaminergic neurons. Later, it was shown that these effects specifically involved the dopamine system.[7]

However, it was not until 1970 that it was actually demonstrated that a variety of antipsychotic drugs (neuroleptics) blocked the function of dopamine receptors, while only a few blocked the actions of noradrenergic receptors.[8] By the mid-1970s it was shown that the antipsychotic action of neuroleptics was correlated with their affinity to dopamine receptors in vitro,[9,10] thereby implicating overactivity of the central dopaminergic system in psychiatric diseases such as schizophrenia.[11,12] In 1979, the idea that dopamine receptors were comprised of two distinct molecular entities, referred to as D1 and D2 receptors, was established[13] and it was quickly noted that neuroleptic activity correlated with affinity at D2 receptors. All of these findings have served to firmly establish the dopaminergic system as one of the central neurotransmitter systems in the brain, with strong influences on cognitive

Recombinant Cell Surface Receptors: Focal Point for Therapeutic Intervention, edited by Michael J. Browne.

function and motor control, not to mention neuroendocrine functions as well as numerous peripheral effects.

The recent discovery of several novel subtypes of dopamine receptors has given rise to the hope that characterization of the interactions of therapeutic drugs with various receptor sites will lead to an understanding of their relevant sites of action. Enhanced antipsychotic activity with a minimized side effect profile, for example, might then be achieved by development of agents exhibiting high selectivity for the relevant receptor site.

2.2 MOLECULAR BIOLOGY OF DOPAMINE RECEPTOR SUBTYPES

Following the development of techniques for analyzing neurotransmitter receptors, it was determined that dopamine acted through two distinct receptor molecules, termed D1 and D2.[13] These two receptors could be distinguished by their pharmacological characteristics, their effects on intracellular second messenger systems, and their different tissue distributions. Initially, D1 receptors were characterized by their lower affinity for dopamine and their ability to activate the enzyme adenylyl cyclase, while D2 receptors were characterized by their higher affinity for dopamine, and their apparent lack of effect on adenylyl cyclase.[13] It was later demonstrated that D2 receptors do in fact affect adenylyl cyclase activity by inhibiting adenylyl cyclase when the enzyme has been pre-stimulated. Subsequently, many differences between the D1 and D2 receptors have been characterized in terms of pharmacological, biochemical, physiological and behavioral properties.[14] Based on the data available at that time, there was little need to invoke the existence of any additional receptor sites to account for all of the actions of dopamine (although there were proposals for distinctly different D1 and D2 receptors in peripheral tissues as opposed to the central nervous system). It was not surprising when the first dopamine receptors to be cloned corresponded precisely to the classically defined D1 and D2 receptors.[15-19] This situation did not last long, however, as these reports were quickly followed by the identification of three additional receptor genes[20-26] as well as two isoforms of the prototypical D2 receptor which are generated by alternative mRNA splicing.[27-38] Interestingly, all of the subtypes of dopamine receptors appear to fall into two distinct families corresponding to the original division of distinct D1 and D2 receptor subtypes.

2.3 GENERAL STRUCTURAL FEATURES OF DOPAMINE RECEPTORS

The determination of the primary sequence of the dopamine receptors has revealed many interesting features of these proteins. First and foremost is their clear membership in the class of proteins known as the G-protein coupled receptor (GPCR) superfamily. The basic structural feature of this protein class consists of seven hydrophobic stretches of approximately 24 amino acids that are believed to represent membrane spanning domains based on the prototypical family member rhodopsin. Another major feature of this protein family is that they initiate signal transduction, wherein the binding of extracellular ligands transmits a signal across the plasma membrane leading to receptor interaction with heterotrimeric GTP-binding proteins (the so-called G-proteins).

The transmembrane topology of these proteins, by analogy with bovine rhodopsin whose structure has been resolved to 9Å,[39] provides an extracellular amino-terminus and an intracellular carboxy-terminus, with alternating intra- and extracellular loops connecting the transmembrane regions. Biochemical studies conducted prior to the cloning of these receptors revealed that they are glycoproteins, and analysis of the cloned genes indicates that they encode proteins having one or more consensus sites for N-linked glycosylation in the extracellular domains. Furthermore, in vitro mutagenesis studies have indicated that ligand binding likely involves the transmembrane domains, particularly the 3rd and 5th transmembrane domains for agonist binding, and the 7th transmembrane domain additionally for antagonist binding (see ref. 40 for review). This concept is particularly important for understanding the pharmacological similarities and differences between dopamine receptor subtypes. Finally, the intracellular loops, particularly the 2nd and 3rd intracellular loops, have been shown to be responsible for coupling with G-proteins, while the C-terminal "tail" has been shown to be important for desensitization and downregulation via receptor internalization.

2.4 THE D1 FAMILY OF DOPAMINE RECEPTORS

The D1 dopamine receptor family consists of at least two members, which have been referred to as D1 and D5 in the case of the human genes, and D1A and D1B in the case of the rat genes. Two additional D1 like receptors have recently been identified in *Xenopus laevis* (D1C)[41] and in chicken (D1D),[42] however, the presence of homologues of these genes in mammalian species has not yet been demonstrated. All of the D1 receptor genes have been identified by techniques which exploit their homology with previously cloned G-protein coupled receptors as a means of isolation. The deduced amino acid sequences of the D1, D5, D1C and D1D receptors are highly homologous (Fig. 2.1), particularly in the transmembrane domains, and all share a common overall topology which includes a short third intracellular loop and long carboxy-terminal (C-terminal) cytoplasmic tail (see Fig. 2.1). At the level of nucleotide and amino-acid sequence these are clearly distinct receptor proteins, however, on a pharmacological level these sites are barely distinguishable with known ligands. Furthermore, all four receptors have been shown to stimulate the activity of adenylyl cyclase.

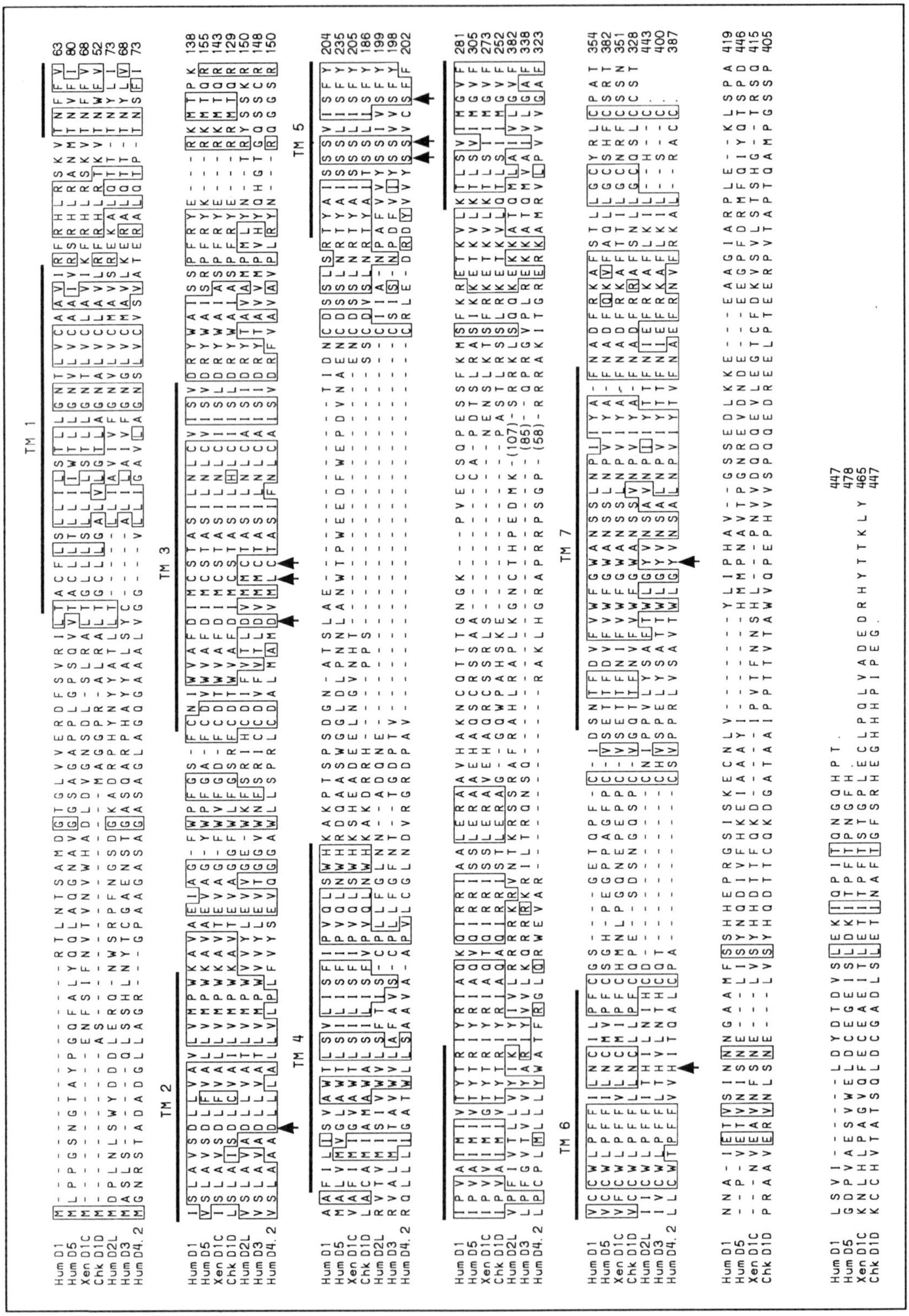

2.4.1 The Dopamine D1 Receptor

The D1 dopamine receptor was initially cloned in 1990 from both rat and human genomic and cDNA libraries by four independent groups.[15-18] The clones obtained from both species possessed an open reading frame of 446 amino acids, and exhibited 91% identity at the amino acid level. Both the human and rat genes lack introns in their coding regions, and the human gene has been localized to chromosome 5.[15,17,18] The D1 receptor contains two potential extracellular N-linked glycosylation sites, one on the amino-terminal tail, and the other on the second extracellular loop. There exists one consensus phosphorylation site for cAMP dependent protein kinase in the third cytoplasmic loop, and a conserved cysteine residue in the C-terminal tail which has been shown to serve as a site for palmitoylation.[43] The C-terminal tail also contains numerous serine and threonine residues which could serve as additional sites for regulatory phosphorylation.

Transient or stable heterologous expression of the D1 receptor clone produces a receptor protein which exhibits the pharmacological and functional characteristics classically associated with the D1 receptor found in tissue preparations.[15-18] The prototypical D1 selective antagonists [^{3}H]-SCH 23390 or [^{125}I]-SCH 23982 exhibit saturable, high affinity binding to the recombinant receptor with appropriate affinity and pharmacological specificity. Functionally, the recombinant receptors can be shown to mediate activation of adenylyl cyclase with a pharmacological profile identical to that described for endogenous receptor expressing systems.

Using both Northern blot and in situ hybridization analyses, D1 receptor mRNA has been shown to be predominantly localized to the striatum and nucleus accumbens, with lesser amounts seen in the olfactory tubercle and cortex,[15-18] correlating well with data based on D1 receptor binding activity[44] and dopamine-stimulated adenylyl cyclase activity. Within the caudate putamen, about 50% of the medium-sized neurons appeared to express the D1 receptor, as assessed by in situ hybridization.[45,46] A small number of large, presumably cholinergic interneurons may also express low levels of D1 receptor mRNA.[47] Current evidence indicates that many of the medium-sized neurons which express high levels of D1 receptor mRNA also express substance P, and belong to the striatal-nigral projection system.[46,47]

Fig. 2.1. (opposite page) Alignment of the amino acid sequence of cloned dopamine receptors. Amino acid residues which are identical in at least four receptor subtypes are boxed. Bars above the sequences indicate the putative transmembrane (TM) regions. Arrows below the sequence indicate highly conserved residues and those which have been studied by in vitro mutagenesis as discussed in the text. A number of residues have been deleted between TM 5 and TM 6 (the third cytoplasmic loop) as indicated in the D2, D3 and D4 receptors to conserve space. The GenBank ascension numbers for the sequences used are: hum D1, X55758;[17] hum D5, M67439;[21] xen D1C, U07865;[41] chk D1D, L36879;[42] hum D2, M29066;[31] hum D3, L20469;[82] hum D4, X58497.[26]

2.4.2 The Dopamine D5 Receptor

The second member of the D1 receptor family, the D5 or D1B receptor, was cloned in 1991 from rat and human genomic DNA libraries by means of homology with the human D1 or 5-HT1A receptors.[20-24] This clone contained an intronless open reading frame of 477 deduced amino acids which was 50% homologous to the D1 receptor overall, and up to 80% identical in the transmembrane regions. As previously noted, the overall topology of the D5 receptor was very similar to that proposed for the D1 receptor, with a short third cytoplasmic loop and a long C-terminal cytoplasmic tail. Additionally, the D5 receptor contains two potential N-linked glycosylation sites, one in the N-terminus and the other in the second extracellular loop, a protein-kinase A consensus phosphorylation site in the third cytoplasmic loop, and a conserved cysteine residue in the C-terminal tail.

Transfection of the D5 receptor clone into mammalian cells resulted in the expression of high affinity, saturable [^{3}H]-SCH 23390 binding, with an affinity virtually identical to that seen with the D1 receptor. Indeed, inhibition of [^{3}H]-SCH 23390 binding by a variety of dopaminergic agents reveals affinities which differ little from the D1 receptor binding affinities. However, a notable exception is the affinity for the endogenous agonist dopamine, which binds to the D5 receptor with a 10-fold higher affinity than to the D1 receptor.[20-24] Stimulation of the recombinant D5 receptor by dopamine or other D1 agonists results in the activation of adenylyl cyclase, however, the maximal response achieved in cell lines expressing the D5R is consistently about 25% lower than that seen for D1 receptors expressed at similar levels.[20] While the molecular basis for these differences remains to be elucidated, it is possible that there will be functional consequences such as differential sensitivity to desensitization and downregulation.

Compared to the D1 receptor, the regional distribution of the D5 receptor is more restricted in the CNS, and it is generally expressed at much lower levels. By Northern blot analysis, the areas of highest expression in the human are frontal and temporal cortex, with lower amounts found in hippocampus, caudate, and brain stem.[22,24] In situ hybridization in slices of rat brain indicated that D5 receptor mRNA was localized to the hippocampus and the parafascicular nucleus of the thalamus.[48]

2.4.3 D1C and D1D Dopamine Receptors

Two novel receptors belonging to the D1 subfamily have recently been described. The D_{1C} receptor was cloned from a *Xenopus laevis* genomic library by screening with the human D1 and D5 sequences.[41] Besides the *Xenopus* homologs of the D1 and D5 receptors (termed *Xen* D_{1A} and D_{1B}), an additional clone was isolated whose deduced amino acid sequence exhibited equivalent (≈55%) homology with both mammalian D1 receptors, and was thus termed *Xen* D_{1C}. When the

D_{1C} clone was expressed transiently in COS-7 cells, it exhibited a pharmacological profile intermediate between the *Xen* D_{1A} and *Xen* D_{1B} clones. For example, the agonists dopamine and 2-amino-6,7,-dihydroxytetralin (ADTN), which bind to the *Xen* D_{1B} with ≈10-fold higher affinity than to the *Xen* D_{1A} (as in the mammalian D1 receptors), bind to the *Xen* D_{1C} with intermediate affinities.[41] Most other agonists exhibited slightly higher affinity for the *Xen* D_{1C} than for either *Xen* D_{1A} or *Xen* D_{1B}. Similarly, in the case of (+) butaclamol and α-flupentixol, two antagonists which bind the *Xen* D_{1A} with ≈10-fold higher affinity than *Xen* D_{1B}, the affinity for *Xen* D_{1C} is intermediate. Other antagonists which do not discriminate between *Xen* D_{1A} and *Xen* D_{1B} likewise do not exhibit any selectivity for the *Xen* D_{1C} subtype.[41] Functionally, the *Xen* D_{1C} is essentially indistinguishable from the other two D1 receptors since this receptor also couples positively to adenylyl cyclase. Transcripts for all three D1 receptor subtypes can be detected in *Xenopus* brain by reverse transcriptase coupled PCR (RT-PCR), however, using this method only transcripts for the *Xen* D_{1B} and *Xen* D_{1C} subtypes could be detected in the *Xenopus* kidney.[41]

Another new D1 receptor subtype has recently been cloned from a chicken (*Gallus domesticus*) genomic library, also using the human D1 receptor sequence as a probe.[42] This new receptor sequence exhibits an overall amino-acid homology of 49% and 46% with the human D1 and D5 receptors , respectively. If only the transmembrane regions are considered this homology rises to 75% and 77%, respectively. When compared to the *Xen* D_{1C} sequence, there is only 55% homology with the chicken D1D receptor, thus indicating that this is indeed a unique receptor gene, and not the avian homolog of the *Xen* D_{1C}.[42] Evidence that the D1D protein encodes a D1-like dopamine receptor is indicated by its ability to bind the D1 antagonist [^{3}H]-SCH 23390 with high affinity, and by a pharmacological profile clearly consistent with that of the human D1 and D5 receptors.[42] Nevertheless, the chicken D1D receptor exhibits some unique pharmacological characteristics. For example, although similar to the D5 and *Xen* D_{1C} in that the D1D binds both dopamine and ADTN with ≈10 fold higher affinity than the D1, the D1D is unique in its 15-25 fold higher affinity for the D1 partial agonist SKF-38393. Other agonists tend to bind to the D1D with slightly higher affinity than to the other D1 subtypes.[42] The affinities of antagonists for the D1D are generally similar to those for the D1 and D5 receptors, with (+)butaclamol exhibiting a lower, D5 like affinity, and flupentixol a somewhat higher, D1-like affinity. Notably, haloperidol exhibits a markedly reduced affinity for the D1D as compared to D1 and D5, while spiroperidol, the classical D2 antagonist, exhibits a relatively high affinity, more similar to that of the human D1 than the D5 (however, it should be noted that the chicken D1A, D1B and D1D receptors all bind spiroperidol with equally high affinity).[42]

Functionally, the D1D receptor resembles the other D1 receptors in activating adenylyl cyclase. Surprisingly though, two compounds typically considered D2 receptor agonists, lisuride and pergolide, were able to activate adenylyl cyclase through the D1D, although at other D1 receptor types these compounds act as antagonists.[42] It seems likely that investigations into the basis for the shift in the character of these compounds would provide insight into the structural features involved in binding of agonists versus antagonists.

The characterization of these two novel members of the D1 receptor family suggests that there is more complexity to this branch of the dopamine receptor family than previously thought. While various lines of evidence over the years suggest that there may in fact be other D1-like receptors present in mammals (see below for further discussion), it remains to be seen whether the D1C and D1D receptors can account for these observations, since genes homologous to the D1C and D1D receptors have yet to be identified in mammalian species.

2.5 THE D2 FAMILY OF DOPAMINE RECEPTORS

Currently the family of D2 receptors consists of four receptor proteins which are the products of three separate genes, one of which is alternatively spliced to generate two different protein products. They are designated the D2, which includes the alternatively spliced D2L (long) and D2S (short), D3, and D4 dopamine receptors. As with the D1 receptors, the grouping of these three genes in a common family of dopamine receptor is based on sequence homologies, pharmacological profiles, and functional activities. The D2-like receptors exhibit approximately 40% amino acid homology overall and 56% within the transmembrane domains (see Fig. 2.1); they all exhibit a D2-like pharmacological profile, although their affinities for specific drugs may vary; and, with the possible exception of the D3 receptor, they appear to couple to the inhibition of adenylyl cyclase through the Gi family of G-proteins.

2.5.1 THE DOPAMINE D2 RECEPTOR

The very first dopamine receptor to be cloned, through low stringency hybridization with the beta-adrenergic receptor sequence, was the D2 dopamine receptor.[19] The D2 cDNA clone isolated in this study consisted of an open reading frame encoding 415 amino acids with the seven hydrophobic stretches typical of G-protein coupled receptors. Three potential N-linked glycosylation sites were found in the putative extracellular N-terminal region. This receptor exhibited a relatively short third cytoplasmic loop and a long cytoplasmic C-terminal tail, features commonly found in receptors which couple to G-proteins acting to inhibit adenylyl cyclase activity. Transfection of this clone into mouse fibroblasts resulted in the expression of high affinity, saturable binding of the D2 antagonist [^{3}H]-spiroperidol,[19] which showed a pharmacological profile indicative of binding to a D2 dopamine receptor based on data obtained with rat striatum membranes.[19]

Soon after the elucidation of the D2 receptor sequence, a second form of the receptor was identified which proved to be generated by alternative splicing of precursor mRNA.[27-38] This splicing event results in the inclusion (D2L form) or exclusion (D2S form) of an 87 base pair exon, resulting in the addition of 29 amino acids to the third cytoplasmic loop as compared to the sequence originally described.[19] This represented the first instance of alternative splicing to be described for G-protein coupled receptors, and raised a number of questions concerning the significance of such variation in the receptor structure. One line of inquiry focused on the possibility that there were regional variations in the expression of one isoform over the other, or that one form represented a post-synaptic and the other a pre-synaptic D2 receptor, while other studies have focused on possible functional or pharmacological differences between the two isoforms.

The regional expression of the long and short isoforms has been studied using oligonucleotide probes which allow the two forms to be differentiated. The results demonstrated that while both forms appear to be expressed in the same brain regions, it is the long isoform which is generally predominant in the brain.[28,30-32,34,36,38,49-52] Furthermore, in situ hybridization studies indicate that the two isoforms do not represent pre-synaptic vs. post-synaptic forms of the D2 receptor.[49]

Functional and pharmacological differences have been examined in cell lines transfected with each isoform. From these studies it is clear that the pharmacology of these isoforms is essentially identical.[27-38] This is not unexpected since the transmembrane regions of the D2L and D2S isoforms are identical. The 29 amino-acid insertion is located in the third cytoplasmic loop, a region thought important for G-protein coupling. Surprisingly though, D2L and D2S were initially found to also be indistinguishable in their coupling to G-proteins and second messenger systems: both isoforms were equivalent in their ability to inhibit adenylyl cyclase activity,[53,54] stimulate PLC activation,[54] mobilize intracellular calcium[55] and activate Na^+/H^+ exchange.[56] However, several recent studies have indicated that these receptor isoforms can exhibit different signaling properties, given the appropriate cellular environment. For example, in the neuroblastoma/glioma hybrid NG108-15, D2L and D2S exhibit differences in signaling[57] which appear to relate to their ability to couple to different classes of heterotrimeric G proteins in this cell line. In another study, it has been shown that the D2 receptor isoforms couple preferentially to distinct forms of the Gi alpha subunit in GH4C1 cells.[53] These are the first studies to convincingly demonstrate signaling differences between the D2 receptor isoforms, albeit in heterologous expression systems. In order to determine the physiological relevance of such coupling differences, these differences will have to be demonstrated in vivo.

2.5.2 The Dopamine D3 Receptor

The D3 receptor was the first of the "new" dopamine receptors to be identified, and was isolated by virtue of its sequence homology with the D2 receptor and other G-protein coupled receptors.[25] At the level of amino acid sequence, it is most homologous to the D2 receptor, exhibiting 75% homology in the transmembrane domains (Fig. 2.1). Other structural features, such as a long third cytoplasmic loop and a short C-terminal domain, reinforce the similarity of the D3 receptor to the previously cloned D2 receptor, and its distinction from the D1 receptor family. Like the D2 receptor, the gene for the D3 receptor contains introns,[25,50,58-60] and alternative splicing events can give rise to D3 receptor isoforms, some of which encode truncated receptor proteins.[60,61] In the mouse, two alternatively spliced forms of the D3 receptor have been discovered which resemble the isoforms of the D2 receptor in that they differ by the presence or absence of a 21 amino acid segment in the third cytoplasmic loop.[58,62] Functional differences between these D3 receptor isoforms remain to be investigated. Interestingly, comparison of the murine and human amino acid sequences in the third cytoplasmic loop reveals that the murine D3 receptor contains 46 additional amino acids which represent 2 short exons not yet described in the human.[58] Since the gene structure for the human D3 receptor has not yet been described, it is not clear whether the published human cDNA sequence represents a short, alternatively spliced isoform, or whether these exonic sequences have been lost during human evolution. By analogy with the D2 receptor, it is likely that such variation in the third cytoplasmic loop will have functional consequences, and thus further characterization of the human D3 receptor gene is necessary for better understanding of the molecular nature of the D3 receptor in the human.

The pharmacology of the D3 receptor expressed in fibroblast cells is similar to that of the D2 receptor, with most antagonists exhibiting slightly higher affinity for the D2 receptor.[25] Interesting exceptions to this profile are the putative autoreceptor selective antagonists AJ76 and UH232, which bind with 3- and 4-fold higher affinity to the D3 receptor, respectively.[25] This suggests that the D3 receptor could represent the elusive pre-synaptic D2-like autoreceptor (see below for further discussion). Agonist binding to the D3 receptor also tends to exhibit higher affinity as compared to the D2 receptor. However, in most cellular systems examined so far, agonist binding to the D3 receptor appears to be unaffected by addition of guanine nucleotides, suggesting a lack of functional interaction with G-proteins. Indeed, in contrast with the D2 receptor, the demonstration of second messenger responses to D3 stimulation has been difficult, but at least one group has been able to demonstrate inhibition of adenylyl-cyclase activity and stimulation of extracellular acidification through D3 receptors expressed in a Chinese Hamster Ovary (CHO) cell line. In this case, however, the response appears to require relatively high levels of D3 expression.[63] Other groups have shown D3 receptor mediated signaling using the D3 receptor transfected into various cell lines; inhibition of calcium currents has been demonstrated in differentiated NG108-15 cells;[64] activation of c-*fos* and stimulation of [^{3}H]-thymidine incorporation in undifferentiated NG108-15 cells;[65,66] and inhibition of dopamine release has been shown in a mesencephalic cell line.[67]

The expression of D3 receptor mRNA in the rat brain only partially overlaps with that of the D2 receptor. In particular, D3 mRNA appears to be localized to regions receiving dopaminergic input from the A_{10} cell group including nucleus accumbens, islands of Calleja, and the bed nucleus of the stria terminalis.[25,68] D3 receptor mRNA can also be found in other limbic brain areas including the olfactory tubercle, hippocampus, mammary nuclei and hypothalamus. Small amounts of D3 receptor message have also been observed in the caudate putamen and certain regions of the cerebral cortex. Importantly, D3 receptor mRNA, as well as D2 mRNA, is expressed in dopaminergic cells in the substantia nigra, indicating that both D3 and D2 function as a presynaptic autoreceptors. The question of co-expression of D2 and D3 receptors in single cells remains to be answered, however. Overall, the pattern of D3 receptor expression suggests that this dopamine receptor subtype may be well positioned to mediate the effects of dopamine on cognitive and emotional functions, and thus could prove to be a relevant target for anti-psychotic drug therapy.

The characteristics of the D3 receptor discussed above have also led to attempts to link mutations in the D3 receptor gene with the etiology of schizophrenia.[69-80] Except for a restriction fragment length polymorphism observed for the D3 gene which apparently is not associated with schizophrenia,[69-72,74-81] no significant mutations in the coding sequence have been noted. On the other hand, one group has reported a loss of D3 receptor mRNA from parietal and motor cortex in postmortem brain samples from schizophrenic patients.[82] However, since these samples have not been well controlled for medication history and other possible confounding variables, it will be necessary for such a finding to be replicated in a larger patient sample.

2.5.3 The Dopamine D4 Receptor

The third member of the D2 receptor subfamily, the D4 receptor, was also identified through the use of the sequence of the D2 receptor. A human neuroblastoma cell line cDNA library was screened with a D2 receptor probe at low stringency and a partial cDNA clone was isolated which exhibited high homology to the D2 receptor transmembrane regions 5, 6 and 7.[26] The corresponding genomic clone, which was isolated from a human genomic library, contained sequences corresponding to the N-terminus through the fourth transmembrane region as well as several introns which interrupted the coding sequence in transmembrane regions 1, 3, 6 and the third cytoplasmic loop. These introns are in positions analogous to the introns found in both the D2 and D3 receptors.[25,26,31,38,50,58,59] The deduced amino acid sequence, based on the combined genomic and cDNA sequences, results in a 387 amino acid open reading frame containing seven hydrophobic regions which correspond to the putative transmembrane regions found in the D2 and D3 receptors. Overall, the homology with the D2 and D3 receptors is 41% and 39%, respectively, and about 56% for both receptors within the transmembrane regions (Fig. 2.1).

Since a full length cDNA clone of the D4 receptor was never isolated (perhaps due to the extremely high G+C nucleotide content of the mRNA), a hybrid gene/cDNA construct was utilized for transfection into COS-7 cells in order to analyze the pharmacological profile of the D4 receptor.[26] The receptor expressed by this construct exhibited specific, saturable [^{3}H]-spiperone binding with an affinity similar to that seen for the D2 and D3 receptors. Examination of the pharmacological profile revealed that the D4 receptor generally bound agonists and antagonists with similar or somewhat lower affinity than did the D2 receptor. However, a notable exception to this profile was observed for the atypical antipsychotic clozapine and its congener octoclothepin, both of which bound to the D4 receptor with approximately 10-fold higher affinity than to the D2 receptor,[26,83] although subsequent studies have reported somewhat lower affinity and selectivity of clozapine for the D4 receptor.[84-87] Nevertheless, this observation is particularly interesting in that the affinity of clozapine for the D4 receptor is close to the reported serum concentration achieved during successful clozapine therapy, leading to the conclusion that therapeutic levels of clozapine will preferentially occupy the D4 receptor as opposed to the D2 receptor.[88] This has led to much interest in the D4 receptor as a potential target for novel antipsychotic drug development, as discussed further below.

The second messenger signaling pathway(s) associated with D4 activation have also been investigated using heterologous expression systems. These studies have demonstrated that the D4 dopamine receptor couples to inhibition of adenylyl cyclase activity,[85,87] stimulation of Na^+/H^+ exchange,[87] and potentiation of stimulated arachidonic acid release.[87] So far, there have been no reports of D4 signaling in native tissues or in vivo.

A unique aspect of the D4 receptor has been discovered by the characterization of D4 receptor sequences in multiple individual human samples.[83] This study has revealed that the D4 sequence is highly polymorphic, involving a 48 base-pair sequence in the third cytoplasmic loop of the receptor.[83] In the original cDNA fragment isolated from a human neuroblastoma cell line, this sequence was repeated twice, but subsequently isolated genomic clones contained fourfold and sevenfold repeats[83] with minor sequence variations. Further screening has led to the identification of individuals bearing up to 10 repeats, represented by 18 different amino acid sequences.[89] The functional significance of this sequence variation is not clear. Although the original report indicated differential effects of sodium on antagonist binding depending on the number of repeat sequences,[83] more extensive characterization of the binding of dopamine to a variety of D4 isoforms, and the effects of GTP analogs on this binding, have revealed little influence of the polymorphic repeat sequence on D4 receptor binding.[90] Nevertheless, other possible effects remain to be investigated, particularly with respect to receptor signaling.

Based initially on its high affinity for clozapine, the D4 receptor has been implicated in the etiology of schizophrenia, and a recent report suggesting that D4 receptor levels are increased in schizophrenic patients has further strengthened this hypothesis.[91] The discovery of extensive genetic polymorphism at the D4 receptor locus opened up the possibility that specific variants of the D4 receptor might be associated with schizophrenic pathology. Numerous groups have subsequently conducted linkage studies examining D4 receptor variants in schizophrenic patients, but the results thus far have found no evidence for linkage of D4 receptor isoforms to genetic susceptibility for schizophrenia in the populations studied.[92-97] On the other hand, a polymorphism involving the presence or absence of a 12 base pair repeat in the N-terminus of the receptor has been shown to be associated with delusional disorders in an Italian population,[98] but the effect of such a change on functional properties of the D4 receptor has yet to be investigated.

2.6 "NEW" DOPAMINE RECEPTORS

A major question which has arisen as a result of the cloning of the novel dopamine receptors is whether or not all members of the dopamine receptor family have been cloned. Despite numerous attempts utilizing a variety of approaches, no molecular evidence for new mammalian dopamine receptors has appeared since the cloning of the D5 receptor in 1991. Nevertheless, some functional[99] and behavioral[100,101] evidence suggests that other dopamine receptors may exist. The strongest case exists for a D1-like receptor which is coupled to phospholipase C (PLC) activation, as opposed to adenylyl-cyclase activation. The first evidence of this activity was found in the kidney, where D1 selective agonists were shown to activate PLC independently of adenylyl-cyclase activation.[102,103] Further indications of this receptor activity, and its presence in brain, was obtained by injection of mRNA prepared from rat striatum into *Xenopus laevis* oocytes.[99] This resulted in the expression of dopamine mediated Ca^{2+} mobilization (a result of PLC activation), which had the characteristics of a D1-like receptor.[99] Additional, direct evidence for activation of PLC by D1 agonists has been obtained in slices of rat striatum.[104] Pharmacological characterization of this response revealed a profile which is not consistent with activity at either the cloned D1 or D5 receptor.[105-107] Furthermore, in rat amygdala where D1 receptors are expressed but do not activate adenylyl cyclase, the D1 agonist SKF 38393 induces a strong stimulation of PLC activity. These data all point to the existence of a unique "D1-like" receptor. However, the lack of a preparation containing "pure" receptor has hindered further biochemical and pharmacological definition of this activity. Thus, until such a novel D1 receptor is actually cloned, the existence of a PLC linked D1 receptor must be deemed hypothetical.

2.7 STRUCTURE-FUNCTION ANALYSIS OF DOPAMINE RECEPTORS

The elucidation of the primary amino acid sequence of G-protein coupled receptors, including dopamine receptors, has provided the tools needed to take the first steps towards understanding the relationship between the structure and the function of these proteins.[108] Structural analysis of rhodopsin and the beta-adrenergic receptors in particular, has allowed determination of the transmembrane topology and indicates that in many cases ligand binding occurs within a "pocket" created between the transmembrane regions. The validation and refinement of hypothetical receptor structures has involved the use of site-directed mutagenesis and molecular modeling to define particular amino-acid residues involved in the formation of the ligand binding "pocket,"[40,108-112] with the aim of eventually being able to custom design ligands with high affinity and selectivity. These studies have focused primarily on determinations of specific residues critical for agonist and antagonist binding, determinations of the orientation of the transmembrane helices with regard to the binding pocket, and the packing or ordering of the transmembrane domains relative to one another.

Comparison of dopamine receptor sequences with those of related receptors,[113] reveals a number of highly conserved amino acids within the transmembrane regions, some of which have been shown to be critically involved in the binding of agonists to adrenergic receptors.[111,112] The conserved aspartic acid residue in transmembrane domain 3 (Fig. 2.1) has been mutated in the D2 dopamine receptor,[114] and as found for the adrenergic receptors, this residue is absolutely required for both agonist and antagonist binding, presumably through hydrogen bonding with the protonated amino group of the affected ligands.[108-110,115,116] Similarly, conserved serine residues in transmembrane domain 5 in both the D1[117,118] and D2 receptors[114,119] (Fig. 2.1) have been shown to be involved in agonist binding, probably through hydrogen bonding to the catechol hydroxy groups of dopamine.[108-110,115,116]

Although these three residues appear to represent the most critical residues for agonist binding (and may be involved in the binding of some antagonists), several studies have identified other residues which may contribute to the formation of the binding pocket. In the D1 receptor, mutation of transmembrane 3 (tm3) residues Cys106 and Ser107 (Fig. 2.1) alter both agonist and antagonist binding affinities,[117] whereas substitution of Trp321 in tm7 (Fig. 2.1) affected only antagonist binding.[117] In tm3 of the D2 receptor, Cys118 (Fig. 2.1) has been shown to be present in the binding pocket.[120,121] Furthermore, using Cys substitution of each residue in transmembrane region 3, and examining the sensitivity of these mutants to sulfhydryl reagents in the presence and absence of ligand, Javitch et al have identified 10 residues which face the binding pocket.[120] Their results confirm the conformation of the transmembrane segment as an alpha helix having 6 turns within the membrane, and confirm the positioning of the conserved Asp residue (Asp114 in D2) in the binding pocket.[120] It should be noted, however, that of the residues in tm3, only the Asp114 is indispensable for ligand binding. A histidine residue (His393 in D2) in tm6 (Fig. 2.1) appears important for the binding of the substituted benzamide class of antagonists, but not other classes of antagonists.[122] Finally, a highly conserved Asp residue in transmembrane region 2 (Asp80 in D2) (Fig. 2.1), while not involved in the binding pocket, has been shown to be critical for the observed regulation of agonist binding by sodium ions and pH, and for agonist mediated inhibition of adenylyl cyclase.[123]

2.8 "KNOCKOUT" OF DOPAMINE RECEPTORS

The development of transgenic animal technologies, and the utilization of homologous recombination to replace wildtype genes with genes altered in vitro has provided an entirely new avenue for analysis of gene/protein function.[124] One approach being utilized to a great extent is that of gene "knockout," where all or part of the target gene is deleted, resulting eventually in an animal completely lacking in the expression of the protein or gene of interest. The requirement of having a cloned mouse gene for the protein of interest is relatively easily met, and thus with the cloning of the dopamine receptor genes, the process of assessing the impact of removal of one or more dopamine receptors has been initiated. One advantage of this technique lies in the absolute specificity of the molecular manipulations. While treatment of an animal with an antagonist drug leads to a temporary, functional knockout of the drug receptor, the specificity of this approach is relatively poor. For example, while there are compounds which provide good selectivity for D1 over D2 receptors (and vice versa), no drugs are yet available which can discriminate between D1 and D5 receptors. Thus, all effects of treatment with a "D1" antagonist must be considered to be a result of the combined blockade of D1 and D5. On the other hand, targeted gene knockout allows the precise deletion of just one of these receptors at a time, hopefully providing for a more specific analysis of the function of a particular receptor subtype. At this time, the results of knocking out the D1 and the D2 receptors have been reported, and work on the remaining three receptors is in progress.

2.8.1 D1 Knockout

The development of the D1 knockout mouse was reported by two groups in 1994.[125-127] As expected, targeted disruption of the D1 receptor gene results in a total loss of D1 receptor expression, a loss which is manifested in a complete loss of D1 antagonist binding,[125-127] and a total lack of responsiveness, in terms of both electrophysiology and behavior, to D1 pharmacological agents.[125,126] The gross anatomy of the homozygous D1 knockout mice (D1 -/-) is essentially normal, although both groups report that the mice are generally smaller than their wild-type littermates. One group attributes this size difference to a difference in food intake, as the D1 -/- mice don't feed well after weaning unless provided with softened chow on the floor of the cage.[127] This group was unable to find any endocrine or metabolic factors otherwise responsible for the size difference.[127] Surprisingly, the basic locomotor activity of the D1 -/- mice is essentially normal, with the exception of a significant decrease in rearing behavior,[127] and a slight hyperactivity in novel environments.[125,126] Interestingly, however, the hyperlocomotor response to cocaine administration is completely absent in the D1 -/- mutants.[125] And, while electrophysiological responses to D1 agents are understandably absent, it was also observed that D2-mediated responses were significantly reduced,[125] supporting the idea that D1 receptors play an enabling role in D2 receptor function.

In contrast to the relatively minimal effects of D1 deletion on locomotion, much greater effects were observed on the neurochemical organization of the striatum in D1 -/- mice.[126] While the expression of D2 family receptors (as detected by [^{3}H]-spiperone binding) was unchanged,[126,127] surprisingly, no signal for the D5 receptor could be detected by [^{3}H]-SCH23390 autoradiography, suggesting a possible downregulation of this receptor site in the absence of the D1 receptor. Several other neurochemical markers were unchanged (enkephalin mRNA,[126,127] DARPP-33 and tyrosine hydroxylase immunoreactivity,[126] and dopamine transporter binding[126]). However, a most striking change was a pronounced decrease in substance P mRNA[127] and immunoreaçtivity[126] and dynorphin immunoreactivity (IR)[126] in the striatum, and particularly in the striosomes, normally sites of high dynorphin expression.[126] Since dynorphin and substance P in the striatum are generally co-expressed with D1 receptors, these results suggest that D1 receptors play an important role in controlling the expression of their genes. Furthermore, it was noted that other markers of striosomal structure, while present, were less well organized than generally seen in wild-type mice.[126] This suggests that the D1 receptor may also play a role in the development and/or maintenance of the striosome/matrix organization of the striatum.

2.8.2 D2 Knockout

The knockout of the D2 receptor has recently been described by Baik et al (ref. 128). Disruption of the D2 receptor gene was achieved by removal of the second exon, which contains the ATG initiation codon. In mice homozygous for this construct (D2 -/-), there was a complete loss of D2 receptor expression as measured by [^{3}H]-spiperone binding, as expected. Interestingly, in heterozygous mice (D2 +/-), D2 receptor binding was reduced by about 50%. The distribution of [^{125}I]-sulpiride binding (which has high affinity for D2 and D3 receptors, but not D4) was examined by in situ ligand autoradiography, and while drastically reduced in D2 -/- mice, residual binding to (presumably) D3 receptors was observed in areas where D3 mRNA has been identified.[129] The binding of the D1 antagonist [^{3}H]-SCH 23390 was found to be unaltered, and it is stated that no changes were observed in the pattern of expression of D3 and D4 receptor genes.[128] These data suggest that D2 receptors play no role in the regulation of expression of other dopamine receptor genes, and that no compensation for the lack of D2 receptors is occurring through other dopamine receptors.

In general, D2 -/- mice were about 15% smaller than their wildtype littermates, and exhibited reduced food and water intake. Although no gross anatomical abnormalities were reported, reduced fertility of the D2 -/- mice was observed, and examination of ovaries and testes revealed a significant reduction in size.[128] Since D2 receptors are expressed in the pituitary where they regulate prolactin synthesis and release, it is suggested that prolactin may play a role in gonadal development.[128]

Despite the lack of effect on expression other dopamine receptors, D2 -/- mice show large changes in other neurochemical systems. Thus enkephalin mRNA (found in D2 expressing cells in wild-type animals) was found to be increased by 40% whereas substance P mRNA (localized to D1 expressing cells in the striatum) was decreased by 15%.[128] In contrast, dynorphin mRNA (also localized to D1 expressing cells in striatum) was unchanged, as was mRNA for tyrosine hydroxylase, suggesting no change in the synthesis of dopamine. Finally, glutamic acid decarboxylase (GAD) mRNA levels were increased in both the cortex (40%) and in the striatum (20%), while unchanged elsewhere. This may reflect disruption of the striato-nigral motor pathway and the associated cortical feedback loop, with alterations in GAD expression (and presumably GABA levels) acting as a partial compensatory mechanism.

Examination of the behavior of D2 -/- mice indicates that the motor pathways in these mice are indeed disrupted. In distinct contrast to the D1 -/- mice described earlier, the D2 -/- mice are strongly impaired in locomotor activity, and even the D2 +/- exhibit some decrements in their locomotor abilities. The D2 -/- mice have abnormal posture and gait, a 74% reduction in locomotion, and complete absence of vertical activity.[128] In addition, these mice exhibit strong cataleptic-like behavior. It is noted that the appearance of these mice very much resembles the effects of D2 antagonist administration, as well as mimicking many of the symptoms seen in Parkinson's disease patients. It is possible that these mice may represent a new animal model for Parkinson's disease, and therefore lead to new insights, and perhaps improved therapies for this disease. However, much further experimentation with these mice is needed in order to determine whether or not any compensatory mechanisms are involved in the final phenotype observed in these animals.

The pronounced effects of D1 or D2 receptor deletions on the neurochemical organization, and presumably, the development of the striatum points up a critical problem with the technology employed in these studies. This is the problem of determining whether behavioral (or indeed any other) effects are truly due to a lack of the target gene expression at the time of testing, or are rather the result of a lack of that gene throughout development. Thus, while D1 and D2 knockout mice may be an important tool for determination of the role of these receptors in developmental processes, their usefulness in defining the role of the D1 or D2 receptor in normal, "adult" functioning is less clear. For example, it is possible that the surprisingly small effect of D1 deletion on normal locomotion may be due to yet to be defined compensatory mechanisms. Due to these problems, determination of the "true" effects of dopamine receptor knockout on brain function will have to await the development of knockout techniques which allow the ablation of a target gene at a time point specified by the investigator. Such "inducible" knockout techniques are currently being developed, and their application to target genes such as the GPCRs will be eagerly awaited.

2.9 THERAPEUTIC IMPLICATIONS OF MULTIPLE DOPAMINE RECEPTOR SUBTYPES

Despite the many years of research into the dopaminergic neurotransmitter system, the two conditions which originally sparked interest in this system, Parkinson's disease and schizophrenia, are still the primary conditions where dopaminergic therapies are indicated. In both of these cases, the treatments currently available leave much to be desired in terms of rate of responsiveness and/or side effect profiles. Because of this situation, the identification of additional dopaminergic receptors has resulted in much hope that more efficacious and better tolerated therapies can be devised.

The classical neuroleptics such as chlorpromazine and haloperidol are effective at alleviating acute 'positive' psychotic symptoms (i.e. hallucinations and paranoia), and they also appear to reduce the number or frequency of psychotic episodes.[130-132] However, these drugs are associated with a high incidence of side effects,[130,133] lack of efficacy in up to 30% of patients, and no benefit for the eventual outcome of the disease. Traditionally, most of these behavioral and neurophysiological effects were believed to be mediated by blockade of D2 receptors in the brain. The identification of multiple D2-like dopamine receptors has required reexamination of this view. Pharmacological studies of the newly cloned D2-like receptors reveal that the classical neuroleptics bind not only to D2 receptors, but also to D3 and D4 receptors with high affinity (Table 2.1). Studies with recombinant receptors have shown that chlorpromazine, the original antipsychotic drug, binds with relatively high affinity to all five dopamine receptor subtypes, and shows some preference for D2 and D3 receptors. Similarly, the neuroleptic haloperidol shows strongest affinity for D2 receptors, but it also binds to D3 and D4 receptors with affinities in the nanomolar range.

Localization studies of the new dopamine receptors have indicated that, in contrast to D2 receptors, both D3 and D4 receptors are expressed at relatively low abundance in the striatum. The localization of D3 and D4 receptor expression in the mesolimbic/mesocortical pathway,[25,26] which controls emotional behavior and is implicated in affective and compulsive behaviors, suggests that neuroleptics selective for these receptor subtypes might have potent antipsychotic activity without producing locomotor side effects.

Based on these pharmacological and anatomical data, a current working hypothesis is that the locomotor or extrapyramidal side effects (EPS) of traditional neuroleptic drugs such as haloperidol are caused by blockade of D2 receptors expressed at high levels in the nigrostriatal pathway which controls voluntary movement, whereas the beneficial therapeutic effects are mediated by the D3 and/or the D4 receptors. This hypothesis is currently driving many efforts aimed at developing novel antipsychotic agents.

In the case of Parkinson's disease, the therapy of choice has changed little since the first successful treatments with L-DOPA were reported in 1969.[4,5] In general, the first years of L-DOPA therapy are reasonably successful with good response rates and relatively few adverse side effects.[134,135] However, with the prolonged treatments required by this chronic condition, many problems arise in terms of fluctuations in efficacy, dyskinesias, freezing, psychological disturbances, and losses of efficacy (requiring increased dosages and consequently increased side effects).[134-136] Several therapeutic strategies have been developed to combat the problems associated with L-DOPA therapy, including additions of anticholinergic drugs, centrally acting monoamine oxidase and catechole-O-methyl transferase inhibitors, and dopaminergic agonists and antagonists (for blockade of peripheral side effects).[134,136]

Four compounds which have in common agonist activity at D2 receptors have been used as adjunct therapies in Parkinson's disease: bromocriptine, lisuride, pergolide and apomorphine.[134,136] Treatment

Table 2.1. Inhibition constants (K_i) for a variety of dopaminergic agonists and antagonists.

K_i Values (nM) Drug		D1	D5	D1C	D1D	D2	D3	D4
Agonists								
Dopamine	High			839	201	474	3.9	28
	Low	2,340	228			1,700	73	450
Apomorphine		680	363	71	18	24	20	4.1
Bromocriptine		700	454		759	5.3	7.4	340
Pergolide		1,363	918		118	21	0.62	
Quinpirole						576	5.1	46
SKF 38393		150	100	72	4	9,560	5,000	1,800
7-OH-DPAT		5,000†				10†	1†	650†
Fenoldopam		1,816	800	5.2	3			321
Antagonists								
Chlorpromazine		73	133	65	70	2.8	6.1	37
Clozapine		141	250	522	652	56	180	9–54¶
Haloperidol		27	48	116	2,600	0.45	9.8	5.1
Raclopride		>72,000#				1.8	3.5	237
Remoxipride						125§	969§	3,690
Risperidone						5†	6.7†	7†
Nemonipride						0.06†	0.3†	0.15†
SCH 23390		0.35	0.3	0.08	0.31	1,000‡		3,560
Spiperone		220	4,500	2,280	91	0.069	0.61	0.08
Sulpiride -S-(-)		36,000	77,270			9.2	25	52
Sulpiride-R-(+)		20,454	28,636			85	422	

Values at each receptor were taken from the following, except as noted in the table: D1 and D5, ref. 22; D1C, ref. 41; D1D, ref. 42; D2 and D3, ref. 137; D4, ref. 26; †, ref. 138; §, ref. 139; ‡, ref. 19; # ref. 15. ¶ Reports of clozapine affinity for the D4 receptor have been highly variable, see ref. 140 for discussion.

with bromocriptine or lisuride appears to reduce the incidence of dyskinesias and fluctuations in L-DOPA efficacy,[134,136] whereas apomorpine can be used to combat frequent, unpredictable "on/off" fluctuations.[136] However, side effects such as nausea, postural hypotension, hallucinations and other psychological disturbances limit their utility.

While all four compounds are high affinity agonists at the D2 receptor, their affinity and activity at other dopamine receptors varies (see Table 2.1). For example, in binding assays bromocriptine is 100 fold selective for cloned D2 and D3 receptors versus D1, D4 and D5 receptors,[88] and while an agonist at the D2 and D3 receptors, it is an antagonist at the D1 receptor (agonism at the D4 and D5 receptors has yet to be investigated). Likewise lisuride is an agonist at the D2 and D3 receptors and an antagonist at the D1 receptor. In contrast both apomorphine and pergolide are agonists at both D1 and D2 receptors with approximately equal potency.[88]

Although not thoroughly investigated, it is highly likely that the lack of selectivity of all of these compounds both compromises their therapeutic efficacy and contributes to their side effect profile. In particular, recent evidence indicates that D2 and D3 receptors have opposing effects on locomotor activity; stimulation of D2 receptors leading to locomotor activation while stimulation of D3 receptors leads to inhibition of locomotor activity (see ref. 65 for discussion). Furthermore, the behavior of D2-knockout mice, wherein specific reduction or ablation of D2 receptors results in Parkinson-like locomotor deficits,[128] suggests that an agonist truly selective for the D2 receptor may exhibit much improved efficacy with a reduction in untoward side effects. Unfortunately, the development of compounds with true selectivity for given receptor subtypes is only now being accomplished, and it will be some time still before the effects of such compounds are known in various model systems.

References

1. Carlsson A, Lindqvist M and Magnusson T. 3,4-dihydroxyphenylalanine and 5-hydroxytryptophan as reserpine antagonists. Nature 1957; 180:1200.
2. Carlsson A, Lindqvist M, Magnusson T et al. On the presence of 3-hydroxytyramine in brain. Science 1958; 127:471.
3. Ehringer H and Hornykiewicz O. Verteilung von Noradrenalin und Dopamin (3-hydroxytyramin) in Gehirn des Menschen und ihr Verhalten bei Erkrankungen des extrapyramidalen Systems. Klin Wochenschr 1960; 38:1236-9.
4. Cotzias GC, Papavasiliou PS and Gellene R. Modification of parkinsonism-chronic treatment with L-dopa. N Engl J Med 1969; 280:337-45.
5. Cotzias GC, Van Woert MH and Schiffer LM. Aromatic amino acids and modification of Parkinsonism. N Engl J Med 1967; 276:374-8.
6. Carlsson A and Lindqvist M. Effect of chlorpromazine or haloperidol on formation of 3-methoxytyramine and normetanephrine in mouse brain. Acta Pharmacol et toxicol. 1963; 20:140-4.
7. Anden NE, Roos BE and Werdinius B. Effects of chlorpromazine, haloperidol and reserpine on the levels of phenolic acids in rabbit corpus striatum. Life Sci 1964; 3:149-58.
8. Anden NE, Butcher SG, Corrodi H et al. Receptor activity and turnover of dopamine and noradrenaline after neuroleptics. Eur J Pharmacol 1970; 11:303-14.
9. Seeman P, Lee T, Chan-Wong M et al. Antipsychotic drug doses and neuroleptic/dopamine receptors. Nature 1976; 261:717-9.
10. Creese I, Burt DR and Snyder SH. Dopamine receptor binding predicts clinical and pharmacological potencies of antischizophrenic drugs. Science 1976; 192:481-3.
11. Carlsson A. The current status of the dopamine hypothesis of schizophrenia. Neuropsychopharmacol 1988; 1:179-86.

12. Seeman P. Dopamine receptors and the dopamine hypothesis of schizophrenia. Synapse 1987; 1:133-52.
13. Kebabian JW and Calne DB. Multiple receptors for dopamine. Nature 1979; 277:93-6.
14. Creese I and Fraser CM. Dopamine receptors. In: Creese I and Fraser CM, ed. Receptor Biochemistry and Methodology. 8. New York: Alan R. Liss, 1987:1.
15. Dearry A, Gingrich JA, Falardeau P et al. Molecular cloning and expression of the gene for a human D1 dopamine receptor. Nature 1990; 347:72-6.
16. Monsma FJ, Jr., Mahan LC, McVittie LD et al. Molecular cloning and expression of a D1 dopamine receptor linked to adenylyl cyclase activation. Proc Natl Acad Sci USA 1990; 87:6723-7.
17. Sunahara RK, Niznik HB, Weiner DM et al. Human dopamine D1 receptor encoded by an intronless gene on chromosome 5. Nature 1990; 347:80-3.
18. Zhou Q-Y, Grandy DK, Thambi L et al. Cloning and expression of human and rat D1 dopamine receptors. Nature 1990; 347:76-80.
19. Bunzow JR, Van TH, Grandy DK et al. Cloning and expression of a rat D2 dopamine receptor cDNA. Nature 1988; 336:783-7.
20. Jarvie KR, Tiberi M, Silvia C et al. Molecular cloning, stable expression and desensitization of the human dopamine D1B/D5 receptor. J Rec Res 1993; 13:573-90.
21. Grandy DK, Zhang Y, Bouvier C et al. Multiple human D5 dopamine receptor genes: a functional receptor and two pseudogenes. Proc Natl Acad Sci USA 1991; 88:9175-9.
22. Sunahara RK, Guan H-C, O'Dowd BF et al. Cloning of the gene for a human dopamine D5 receptor with higher affinity for dopamine than D1. Nature 1991; 350:614-9.
23. Tiberi M, Jarvie KR, Silvia C et al. Cloning, molecular characterization, and chromosomal assignment of a gene encoding a second D1 dopamine receptor subtype: Differential expression pattern in rat brain compared with the D1A receptor. Proc Natl Acad Sci USA 1991; 88:7491-5.
24. Weinshank RL, Adham N, Macchi M et al. Molecular cloning and characterization of a high affinity dopamine receptor (D1β) and its pseudogene. The Journal of Biological Chemistry 1991; 266:22427-35.
25. Sokoloff P, Giros B, Martres M-P et al. Molecular cloning and characterization of a novel dopamine receptor (D3) as a target for neuroleptics. Nature 1990; 347:146-51.
26. Van Tol HHM, Bunzow JR, Guan H-C et al. Cloning of the gene for a human dopamine D4 receptor with high affinity for the antipsychotic clozapine. Nature 1991; 350:610-4.
27. Chio CL, Hess GF, Graham RS et al. A second molecular form of D2 dopamine receptor in rat and bovine caudate nucleus. Nature 1990; 343:266-9.
28. Dal Toso R, Sommer B, Ewert M et al. The dopamine D2 receptor: two molecular forms generated by alternative splicing. EMBO J 1989; 8:4025-34.

29. Eidne KA, Taylor PL, Zabavnik J et al. D2 receptor, a missing exon. Nature 1989; 342:865.
30. Gandelman KY, Harmon S, Todd RD et al. Analysis of the structure and expression of the human dopamine D2A receptor gene. J Neurochem 1991; 56:1024-9.
31. Grandy DK, Marchionni MA, Makam H et al. Cloning of the cDNA and gene for a human D2 dopamine receptor. Proc Natl Acad Sci USA 1989; 86:9762-6.
32. Giros B, Sokoloff P, Martres MP et al. Alternative splicing directs the expression of two D2 dopamine receptor isoforms. Nature 1989; 342:923-6.
33. Miller JC, Wang Y and Filer D. Identification by sequence analysis of a second rat brain cDNA encoding the dopamine (D2) receptor. Biochem Biophys Res Commun 1990; 166:109-12.
34. Monsma FJ, Jr., McVittie LD, Gerfen CR et al. Multiple D2 dopamine receptors produced by alternative RNA splicing. Nature 1989; 342:926-9.
35. Montmayeur JP and Borrelli E. Transcription mediated by a cAMP-responsive promoter element is reduced upon activation of dopamine D2 receptors. Proc Natl Acad Sci USA 1991; 88:3135-9.
36. Mack KJ, O'Malley KL and Todd RD. Differential expression of dopaminergic D2 receptor messenger RNAs during development. Brain Res Dev Brain Res 1991; 59:249-51.
37. O'Dowd BF, Nguyen T, Tirpak A et al. Cloning of two additional catecholamine receptors from rat brain. FEBS Lett 1990; 262:8-12.
38. O'Malley KL, Mack KJ, Gandelman KY et al. Organization and expression of the rat D2A receptor gene: identification of alternative transcripts and a variant donor splice site. Biochemistry 1990; 29:1367-71.
39. Schertler GFX, Villa C and Henderson R. Projection structure of rhodopsin. Nature 1993; 362:770-2.
40. Savarese TM and Fraser CM. *In vitro* mutagenesis and the search for structure-function relationships among G protein-coupled receptors. Biochem J 1992; 283:1-19.
41. Sugamori RK, Demchyshyn LL, Chung M et al. D1A, D1B, and D1C dopamine receptors from Xenopus laevis. Proc Natl Acad Sci USA 1994; 91:10536-40.
42. Demchyshyn LL, Sugamori KS, Lee FJS et al. The dopamine D1D receptor. J Biol Chem 1995; 270:4005-12.
43. Ng GY, Mouillac B, George SR et al. Desensitization, phosphorylation and palmitoylation of the human dopamine D1 receptor. Eur J Pharmacol 1994; 267:7-19.
44. Mansour A, Meador WJ, Zhou QY et al. A comparison of D1 receptor binding and mRNA in rat brain using receptor autoradiographic and in situ hybridization techniques. Neuroscience 1991; 45:359-71.
45. Meador WJ, Mansour A, Civelli O et al. Distribution of D2 dopamine receptor mRNA in the primate brain. Prog Neuropsychopharmacol Biol Psychiatry 1991; 15:885-93.

46. Gerfen CR, Engber TM, Mahan LC et al. D1 and D2 dopamine receptor-regulated gene expression of striatonigral and striatopallidal neurons. Science 1990; 250:1429-32.
47. Le Moine C, Normand E and Bloch B. Phenotypical characterization of the rat striatal neurons expressing the D1 dopamine receptor gene. Proc Natl Acad Sci USA 1991; 88:4205-9.
48. Meador WJ, Mansour A, Grandy DK et al. Distribution of D5 dopamine receptor mRNA in rat brain. Neurosci Lett 1992; 145:209-12.
49. Le Moine C and Bloch B. Rat striatal and mesencephalic neurons contain the long isoform of the D2 dopamine receptor mRNA. Brain Res Mol Brain Res 1991; 10:283-9.
50. Mack KJ, Todd RD and O'Malley KL. The mouse dopamine D2A receptor gene: sequence homology with the rat and human genes and expression of alternative transcripts. J Neurochem 1991; 57:795-801.
51. Neve KA, Neve RL, Fidel S et al. Increased abundance of alternatively spliced forms of D2 dopamine receptor mRNA after denervation. Proc Natl Acad Sci USA 1991; 88:2802-6.
52. Snyder LA, Roberts JL and Sealfon SC. Distribution of dopamine D2 receptor mRNA splice variants in the rat by solution hybridization/protection assay. Neurosci Lett 1991; 122:37-40.
53. Senogles SE. The D2 dopamine receptor isoforms signal through distinct $G_i\alpha$ proteins to inhibit adenylyl cyclase. J Biol Chem 1994; 269:23120-7.
54. Liu YF, Civelli O, Grandy DK et al. differential sensitivity of the short and long human dopamine D_2 receptor subtypes to protein kinase C. J Neurochem 1992; 59:2311-7.
55. Hayes G, Biden TJ, Selbie LA et al. Structural subtypes of the dopamine D2 receptor are functionally distinct: expression of the cloned D2A and D2B subtypes in a heterologous cell line. Mol Endocrinol 1992; 6:920-6.
56. Neve KA, Kozlowski MR and Rosser MP. Dopamine D2 receptor stimulation of Na^+/H^+ exchange assessed by quantification of extracellular acidification. J Biol Chem 1992; 267:25748-53.
57. Liu L-X, Monsma FJ, Jr., Sibley DR et al. D_{2S} and D_{2L} receptors couple to K^+ currents in NG108-15 cells via different signal transduction pathways. Society for Neuroscience Abstracts 1994; 20:523.
58. Park B-H, Fishburn S, Carmon S et al. Structural organization of the murine D3 dopamine receptor gene. J Neurochem 1995; 64:482-6.
59. Giros B, Martres MP, Sokoloff P et al. Gene cloning of human dopaminergic D3 receptor and identification of its chromosome. C R Acad Sci Iii 1990; 311:501-8.
60. Giros B, Martres MP, Pilon C et al. Shorter variants of the D3 dopamine receptor produced through various patterns of alternative splicing. Biochem Biophys Res Commun 1991; 176:1584-92.
61. Snyder LA, Roberts JL and Sealfon SC. Alternative transcripts of the rat and human dopamine D3 receptor. Biochem Biophys Res Commun 1991; 180:1031-5.
62. Fishburn CS, Belleli D, David C et al. A novel short isoform of the D3

dopamine receptor generated by alternative splicing in the third cytoplasmic loop. J Biol Chem 1993; 268:5872-8.
63. Chio CL, Lajiness ME and Huff RM. Activation of heterologously expressed D3 dopamine receptors: Comparison with D2 dopamine receptors. Mol Pharmacol 1994; 45:51-60.
64. Seabrook GR, Kemp JA, Freedman SB et al. Functional expression of human D_3 dopamine receptors in differentiated neuroblastoma x glioma NG108-15 cells. Br J Pharmacol 1994; 111:391-3.
65. Sautel F, Griffon N, Lévesque D et al. A functional test identifies dopamine agoinsts selective for D3 versus D2 receptors. Neuroreport 1995; 6:329-32.
66. Pilon C, Lévesque D, Dimitriadou V et al. Functional coupling of the human dopamine D_3 receptor in a transfected NG108-15 neuroblastoma-glioma hybrid cell line. Eur J Pharmacol-Mol Pharmacol Sect 1994; 268:129-39.
67. Tang L, Todd RD and O'Malley KL. Dopamine D2 and D3 receptors inhibit dopamine release. J Pharmacol Exp Ther 1994; 270:475-9.
68. Bouthenet ML, Souil E, Martres MP et al. Localization of dopamine D3 receptor mRNA in the rat brain using in situ hybridization histochemistry: comparison with dopamine D2 receptor mRNA. Brain Res 1991; 564:203-19.
69. Di Bella D, Catalano M, Strukel A et al. Distribution of the MscI polymorphism of the dopamine D3 receptor in an Italian psychotic population. Psychiatr Genet 1994; 4:39-42.
70. Laurent C, Savoye C, Samolyk D et al. Homozygosity at the dopamine D3 receptor locus is not associated with schizophrenia. J Med Genet 1994; 31:260
71. Mant R, Williams J, Asherson P et al. Relationship between homozygosity at the dopamine D3 receptor gene and schizophrenia. Am J Med Genet 1994; 54:21-6.
72. Nimgaonkar VL, Zhang XR, Caldwell JG et al. Association study of schizophrenia with dopamine D3 receptor gene polymorphisms: probable effects of family history of schizophrenia? Am J Med Genet 1993; 48:214-7.
73. Nanko S, Sasaki T, Fukuda R et al. A study of the association between schizophrenia and the dopamine D3 receptor gene. Hum Genet 1993; 92:336-8.
74. Nothen MM, Cichon S, Propping P et al. Excess of homozygosity at the dopamine D3 receptor gene in schizophrenia not confirmed. J Med Genet 1993; 30:708-9.
75. Sabate O, Campion D, d'Amato T et al. Failure to find evidence for linkage or association between the dopamine D3 receptor gene and schizophrenia. Am J Psychiatry 1994; 151:107-11.
76. Morel RT. Association between schizophrenia and homozygosity at the dopamine D3 receptor gene [letter]. J Med Genet 1993; 30:708-9.
77. Yang L, Li T, Wiese C et al. No association between schizophrenia and homozygosity at the D3 dopamine receptor gene. Am J Med Genet 1993; 48:83-6.

78. Wiese C, Lannfelt L, Kristbjarnarson H et al. No evidence of linkage between schizophrenia and D3 dopamine receptor gene locus in Icelandic pedigrees. Psychiatry Res 1993; 46:69-78.
79. Jonsson E, Lannfelt L, Sokoloff P et al. Lack of association between schizophrenia and alleles in the dopamine D3 receptor gene. Acta Psychiatr Scand 1993; 87:345-9.
80. Crocq MA, Mant R, Asherson P et al. Association between schizophrenia and homozygosity at the dopamine D3 receptor gene. J Med Genet 1992; 29:858-60.
81. Nanko S, Hattori M, Ueki A et al. Dopamine D3 and D4 receptor gene polymorphisms and Parkinson's disease. Lancet 1993; 342:250
82. Schmauss C, Haroutunian V, Davis KL et al. Selective loss of dopamine D3-type receptor mRNA expression in parietal and motor cortices of patients with chronic schizophrenia. Proc Natl Acad Sci, USA 1993; 90:8942-6.
83. Van Tol HHM, Wu CM, Guan H-C et al. Multiple dopamine D4 receptor variants in the human population. Nature 1992; 358:149-52.
84. Mills A, Allet B, Bernard A et al. Expression and characterization of human D4 dopamine receptors in baculovirus-infected insect cells. FEBS Lett 1993; 320:130-4.
85. McHale M, Coldwell MC, Herrity N et al. Expression and functional characterisation of a synthetic version of the human D4 dopamine receptor in a stable human cell line. FEBS Lett 1994; 345:147-50.
86. Lahti RA, Evans DL, Stratman NC et al. Dopamine D4 versus D2 receptor selectivity of dopamine receptor antagonists: possible therapeutic implications. Eur J Pharmacol 1993; 236:483-6.
87. Chio CL, Drong RF, Riley DT et al. D4 dopamine receptor-mediated signaling events determined in transfected Chinese hamster ovary cells. J Biol Chem 1994; 269:11813-9.
88. Seeman P. Dopamine receptor sequences: therapeutic levels of neuroleptics occupy D2 receptors, clozapine occupies D4. Neuropsychopharm 1992; 7:261-84.
89. Lichter JB, Barr CL, Kennedy JL et al. A hypervariable segment in the human dopamine receptor D4 (DRD4) gene. Hum Mol Genet 1993; 2:767-73.
90. Asghari V, Schoots O, van Kats S et al. Dopamine D4 receptor repeat: analysis of different native and mutant forms of the human and rat genes. Mol Pharmacol 1994; 46:364-73.
91. Seeman P, Guan HC and Van Tol H. Dopamine D4 receptors elevated in schizophrenia. Nature 1993; 365:441-5.
92. Barr CL, Kennedy JL, Lichter JB et al. Alleles at the dopamine D4 receptor locus do not contribute to the genetic susceptibility to schizophrenia in a large Swedish kindred. Am J Med Genet 1993; 48:218-22.
93. Campion D, d'Amato T, Bastard C et al. Genetic study of dopamine D1, D2, and D4 receptors in schizophrenia. Psychiatry Res 1994; 51:215-30.

94. Coon H, Byerley W, Holik J et al. Linkage analysis of schizophrenia with five dopamine receptor genes in nine pedigrees. Am J Hum Genet 1993; 52:327-34.
95. Macciardi F, Petronis A, Van Tol H et al. Analysis of the D4 dopamine receptor gene variant in an Italian schizophrenia kindred. Arch Gen Psychiatry 1994; 51:288-93.
96. Shaikh S, Gill M, Owen M et al. Failure to find linkage between a functional polymorphism in the dopamine D4 receptor gene and schizophrenia. Am J Med Genet 1994; 54:8-11.
97. Sommer SS, Lind TJ, Heston LL et al. Dopamine D4 receptor variants in unrelated schizophrenic cases and controls. Am J Med Genet 1993; 48:90-3.
98. Catalano M, Nobile M, Novelli E et al. Distribution of a novel mutation in the first exon of the human dopamine D4 receptor gene in psychotic patients. Biol Psychiatry 1993; 34:459-64.
99. Mahan LC, Burch RM, Monsma JFJ. et al. Expression of striatal D1 dopamine receptors coupled to inositol phosphate production and Ca^{2+} mobilization in *Xenopus* oocytes. Proc Natl Acad Sci USA 1990; 87:2196-200.
100. Daly SA and Waddington JL. Behavioural evidence for "D1-like" dopamine receptor subtypes in rat brain using the new isochroman agonist A 68930 and isoquinoline antagonist BW 737C. Psychopharmacology-Berl 1993; 113:45-50.
101. Downes RP and Waddington JL. Grooming and vacuous chewing by SK&F 83959, an agonist of dopamine 'D1-like" receptors that inhibits dopamine-sensitive adenylyl cyclase. Eur J Pharmacol 1993; 234:135-6.
102. Felder CC, Blecher M and Jose PA. Dopamine-1-mediated stimulation of phospholipase C activity in rat renal cortical membranes. J Biol Chem 1989; 264:8739-45.
103. Felder CC, Jose PA and Axelrod J. The dopamine-1 agonist, SKF 82526, stimulates phospholipase-C activity independent of adenylate cyclase. J Pharmacol Exp Therap 1989; 248:171-5.
104. Undie AS and Friedman E. Dopamine stimulates phosphoinositide metabolism in rat brain. Neurosci Res Comm 1990; 6:69-75.
105. Undie AS and Friedman E. Stimulation of a dopamine D1 receptor enhances inositol phosphate formation in rat brain. J Pharmacol Exp Therap 1990; 253:987-92.
106. Undie SA and Friedman E. Selective dopaminergic mechanism of dopamine and SKF 38393 stimulation of inositol phosphate formation in rat brain. Eur J Pharmacol 1992; 226:297-302.
107. Undie AS, Weinstock J, Sarau HM et al. Evidence for a distinct D1-like dopamine receptor that couples to activation of phosphoinositide metabolism in brain. J Neurochem 1994; 62:2045-8.
108. Hibert MF, Trumpp-Kallmeyer S, Bruinvels A et al. Three-dimensional models of neurotransmitter G-binding protein-coupled receptors. Mol Pharmacol 1991; 40:8-15.

109. Teeter MM, Froimowitz M, Stec B et al. Homology modeling of the dopamine D2 receptor and its testing by docking of agonists and tricyclic antagonists. J Med Chem 1994; 37:2874-88.
110. Trump-Kallmeyer S, Hoflack J, Bruinvels A et al. Modeling of G-protein-coupled receptors: Application to dopamine, adrenaline, serotonin, acetylcholine, and mammalian opsin receptors. J Med Chem 1992; 35:3448-62.
111. Strader C, Candelore MR, Hill WS et al. Identification of two serine residues involved in agoinst activation of the beta-adrenergic receptor. J Biol Chem 1989; 264:13572-8.
112. Strader CD, Sigal IS, Candelore MR et al. Conserved aspartic acid residues 79 and 113 of the beta-adrenergic receptor have different roles in receptor function. J Biol Chem 1988; 263:10267-71.
113. Probst WC, Snyder LA, Schuster DI et al. Sequence alignment of the G-protein coupled receptor superfamily. DNA and Cell Biol 1992; 11:1-20.
114. Mansour A, Meng F, Meador-Woodruff JH et al. Site-directed mutagenesis of the human dopamine D2 receptor. Eur J Pharmacol 1992; 1992:205-14.
115. Livingstone CD, Strange PG and Naylor LH. Molecular modelling of D2-like dopamine receptors. Biochem J 1992; 287:277-82.
116. Livingstone CD, Strange PG and Naylor LH. Three dimensional models of the rat D2, D3 and D4 dopamine receptors. Biochem Soc Trans 1992; 20:148S.
117. Tomic M, Seeman P, George SR et al. Dopamine D1 receptor mutagenesis: Role of amino acids in agonist and antagonist binding. Biochem Biophys Res Comm 1993; 191:1020-7.
118. Pollock NJ, Manelli AM, Hutchins CW et al. Serine mutations in transmembrane V of the dopamine D1 receptor affect ligand interactions and receptor activation. J Biol Chem 1992; 267:17780-6.
119. Cox BA, Henningsen RA, Spanoyannis A et al. Contributions of conserved serine residues to the interactions of ligands with dopamine D2 receptors. J Neurochem 1992; 59:627-35.
120. Javitch JA, Fu D, Chen J et al. Mapping the binding-site crevice of the dopamine D2 receptor by the substituted-cysteine accessibility method. Neuron 1995; 14:825-31.
121. Javitch JA, Li X, Kaback J et al. A cysteine residue in the third membrane-spanning segment of the human D2 dopamine receptor is exposed in the binding-site crevice. Proc Natl Acad Sci, USA 1994; 91:10355-9.
122. Woodward R, Daniell SJ, Strange PG et al. Structural studies on D2 dopamine receptors: Mutation of a histidine residue specifically affects the binding of a subgroup of substituted benzamide drugs. J Neurochem 1994; 62:1664-9.
123. Neve KA, Cox BA, Henningsen RA et al. Pivotal role for aspartate-80 in the regulation of dopamine D2 receptor affinity for drugs and inhibition of adenylyl cyclase. Mol Pharmacol 1991; 39:733-9.

124. Capecchi MR. Altering the genome by homologous recombination. Science 1989; 244:1288-92.
125. Xu M, Hu X-T, Cooper DC et al. Elimination of cocaine-induced hyperactivity and dopamine-mediated neurophysiological effect in dopamine D1 receptor mutant mice. Cell 1994; 79:945-55.
126. Xu M, Moratalla R, Gold LH et al. Dopamine D1 receptor mutant mice are deficient in striatal expression of dynorphin and in dopamine-mediated behavioral responses. Cell 1994; 79:729-42.
127. Drago J, Gerfen CR, Lachowicz JE et al. Altered striatal function in a mutant mouse lacking D1A dopamine receptors. Proc Natl Acad Sci USA 1994; 91:12564-8.
128. Baik J-H, Picetti R, Saiardi A et al. Parkinsonian-like locomotor impairment in mice lacking dopamine D2 receptors. Nature 1995; 377:424-8.
129. Levesque D, Diaz J, Pilon C et al. Identification, characterization, and localization of the dopamine D3 receptor in rat brain using [^{3}H]7-OH-DPAT. Proc Natl Acad Sci USA 1992; 89:8155-9.
130. Kane JM. The current status of neuroleptic therapy. J Clin Psych 1989; 50:322-8.
131. Cole JO, Goldberg SC and Klerman GL. Phenothiazine treatment in acute schizophrenia. Arch Gen Psych 1966; 10:246-61.
132. Davis JM, Schaffer CB, Killian GA et al. Important issues in the drug treatment of schizophrenia. Schiz Bull 1980; 6:70-87.
133. Ayd FR, Jr. A survey of drug-induced extrapyramidal reactions. J Am Med Assoc 1961; 175:1054-60.
134. Kopin IJ. The pharmacology of parkinson's disease therapy; an update. Ann Rev Pharmacol Toxicol 1993; 32:467-95.
135. Riederer P, Lange KW and Youdim MBH. Recent advances in pharmacological therapy of Parkinson's disease. In: Narabayashi H, Nagatsu T, Yangisawa N and Mizuno Y, ed. Advances in Neurology. 60. New York: Raven Press, 1993.
136. Coleman RJ. Current drug therapy for parkinson's disease. Drugs and Aging 1992; 2:112-24.
137. Sokoloff P, Andrieux M, Besançon R et al. Pharmacology of human dopamine D3 receptor expressed in a mammalian cell line: comparison with the D2 receptor. Eur J Pharmacol - Mol Pharmacol Sect 1992; 225:331-7.
138. Seeman P and Van Tol HHM. Dopamine receptor pharmacology. Trends Pharmacol Sci 1994; 15:264-70.
139. Mohell N, Sällemark M, Rosqvist S et al. Binding characteristics of remoxipride and its metabolites to dopamine D2 and D3 receptors. Eur J Pharm 1993; 238:121-5.
140. Hartmann D, Monsma F and Civelli O. Interaction of antipsychotic drugs with dopamine receptor subtypes. In: Csernansky JG, ed. Handbook of Experimental Pharmacology, "Antipsychotics". in press.

CHAPTER 3

Molecular Biology of Opiate Receptors

Melanie Tallent, Allan Blake and Terry Reisine

3.1 INTRODUCTION

The opiates are a class of drug which are used extensively in the treatment of pain.[1,2] One of the most widely used and effective analgesics is morphine. Morphine has the desirable property of alleviating pain without the loss of consciousness.[1,2] Codeine, a derivative of morphine, is another commonly used opiate which is an effective cough suppressant as well as an analgesic. Another opiate is methadone, which is used in the treatment of heroin addiction, as it is more water-soluble than morphine and thus can be delivered in orally tapered dosages. Thus the opiates have a range of important clinical uses.

While opiates have a number of clinical uses, they also have side-effects. They cause respiratory depression through their actions in brain stem nuclei.[1] In addition they can cause nausea through their actions in emesis centers of the brain and can effect gut motility through actions on the gastrointestinal system. Chronic use of opiates also causes the development of tolerance and morphine and its derivatives have high abuse potential and are addictive.

The effects of opiates such as morphine are highly selective and in general very potent. The stereoselective actions of opiates are blocked by antagonists such as naloxone. These properties suggested that opiates may act via specific receptors to induce their effects rather than through non-selective membrane destabilizing actions as ascribed to general anesthetics. In the early 1970s direct biochemical evidence was provided that morphine and other opiates interacted with membrane associated receptors. Three major receptors were identified named mu, kappa and delta.[3-9] These receptors have distinct pharmacological characteristics and are localized to regions of the nervous system known to mediate the pharmacological effects of opiates.[10]

Recombinant Cell Surface Receptors: Focal Point for Therapeutic Intervention,
edited by Michael J. Browne.

The identification of opiate receptors raised the question of the identity of their endogenous ligands, since morphine is a plant alkaloid and therefore is not produced in the body. In the mid 1970s Hans Kosterlitz and his associate John Hughes identified the enkephalins, which are peptides expressed in the nervous system which selectively interact with opiate receptors. Two other families of opiate peptides, the endorphins and dynorphins were identified following this initial discovery.

These three families of peptides are encoded by different genes and are synthesized in distinct regions of the nervous system. The enkephalins and endorphins have high affinity for the mu and delta receptors but exhibit very low affinity for kappa receptors. In contrast, dynorphin A has high affinity for kappa receptors and lower affinity for mu and kappa receptors. Thus, the endogenous opioids have different selectivities for the opioid receptor class.

3.2 THE OPIATE RECEPTORS

The three major classes of opiate receptors have distinct pharmacologies and drug selectivities.[9-19] The mu receptor has highest affinity for morphine and other clinically used opiates such as codeine, methadone and fentanyl (see Table 3.1). Highly selective agonists at the mu receptor are the synthetic peptides DAMGO and dermorphin. The actions of opiates at mu receptors can be blocked by the non-selective

Table 3.1. Drug selectivities of opiate receptors

Receptor	Transmitter	Selective Agonist	Selective Antagonist
mu	enkephalin endorphin	morphine methadone fentanyl DAMGO	CTOP naloxonazine
kappa	Dynorphin A	U50,488 spiradoline	nor-BNI
delta	enkephalin endorphin	DPDPE DSLET deltophin II SIOM BW3734U86	naltrindole NTB BNTX

BNTX = 7-benzylidenenaltrexone; CTOP = DPhe-Cys-Tyr-DTrp-Orn-Pen-Thr-NH2; DAMGO = [DAla^2-MePhe^4,Gly-ol^5]enkephalin; DPDPE = [DPen^2 D-Pen^5]enkephalin; DSLET = [DSer^2, DLeu^5]enkephalyl-Thr; NTB = benzofuran analog of naltrindole.

opioid antagonist naloxone. Selective antagonists at the mu receptor include the peptide CTOP and the non-peptide naloxonazine.

Mu receptor agonists are the most potent and effective analgesics.[1,2] However, they also induce severe respiratory depression, nausea, constipation, and are highly addictive when used chronically in an uncontrolled manner.[1] Also, tolerance can occur to the therapeutic actions of mu agonists which can limit their clinical usefulness.

Kappa receptors were originally classified as mediating the psychotomimetic effects of ethylketocyclazocine and its analogs.[1] Stimulants of kappa receptors were found to induce dysphoria rather than the euphoria associated with activation of mu receptors. Kappa agonists are less prone to induce respiratory depression than morphine and have lower abuse potential.

Kappa receptors are distinguished by their high affinity for the agonist U50,488 and its derivatives. A selective antagonist of kappa receptors is nor-BNI. While ethylketocyclazocine has high affinity for kappa receptors, it also binds potently to mu receptors. Its preferential actions at kappa receptors may be related to its full agonism at kappa receptors and its partial agonism at mu receptors.

Kappa agonists can induce analgesia, although not as effectively as morphine. Much of the therapeutic effectiveness of kappa agonists may be via peripheral analgesic mechanisms. As a result, highly selective kappa agonists that do not enter the brain may have particular clinical usefulness since such agents would not be expected to cause those side-effects of kappa agonists—dysphoria, sedation, diuresis, which are mediated by central mechanisms.

The third major class of opiate receptors are the delta receptors. These receptors are distinguished by their high affinity for peptide agonists such as DPDPE, DSLET, and deltorphin II and for the antagonist naltrindole. Recently, selective non-peptide delta receptor agonists, BW3734U86 and SIOM, have been developed which should facilitate further evaluation of the functional role of delta receptors.

Two subtypes of delta receptors have been identified based on their different affinities for agonists and antagonists.[13-15,17,18] The antagonists BNTX and NTB preferentially bind to $delta_1$ and $delta_2$ receptors, respectively. Furthermore, DPDPE preferentially binds to $delta_1$ receptors whereas deltorphin II preferentially labels $delta_2$ receptors. Despite pharmacological evidence of delta receptor subtypes, there is no physical evidence for subtypes as of yet.

3.3 THE CLONING OF THE OPIATE RECEPTORS

Cloning studies provided the first structural information of the opiate receptors[11] (see Fig. 3.1 for sequence). The delta opiate receptor was initially cloned from NG-108 cell libraries by two different research groups.[19,20] The sequence had very little similarity with any other proteins except the somatostatin receptors, of which the similarity was

	5	15	25	35	45
human delta	----------	----------	-MEPAPSAGA	ELQPPLFANA	SDAYPSACPS
human mu	MDSSAAPTNA	SNCTDALAYS	SCSPAPSPGS	WVNLSHLDGN	LSDPCGPNRT
human mu splice	MDSSAAPTNA	SNCTDALAYS	SCSPAPSPGS	WVNLSHLDGD	LSDPCGPNRT
human kappa	--------ME	SPIQIFRGEP	GPTCAPSACL	PPN-------	-SSAWFPGWA

	55	65	75	85	95
human delta	AGANASGPPG	-----ARSAS	SLALAIAITA	LYSAVCAVGL	LGNVLVMFGI
human mu	DLGGRDSLCP	P----TGSP-	SMITAITIMA	LYSIVCVVGL	FGNFLVMYVI
human mu splice	DLGGRDSLCP	P----TGSP-	SMITAITIMA	LYSIVCVVGL	FGNFLVMYVI
human kappa	EPDSNGSAGS	EDAQLEPAHI	SPAIPVIITA	VYSVVFVVGL	VGNSLVMFVI

	105	115	125	135	145
human delta	VRYTKMKTAT	NIYIFNLALA	DALATSTLPF	QSAKYLMETW	PFGELLCKAV
human mu	VRYTKMKTAT	NIYIFNLALA	DALATSTLPF	QSVNYLMGTW	PFGTILCKIV
human mu splice	VRYTKMKTAT	NIYIFNLALA	DALATSTLPF	QSVNYLMGTW	PFGTILCKIV
human kappa	IRYTKMKTAT	NIYIFNLALA	DALVTTTMPF	QSTVYLMNSW	PFGDVLCKIV

	155	165	175	185	195
human delta	LSIDYYNMFT	SIFTLTMMSV	DRYIAVCHPV	KALDFRTPAK	AKLINICIWV
human mu	ISIDYYNMFT	SIFTLCTMSV	DRYIAVCHPV	KALDFRTPRN	AKIINVCNWI
human mu splice	ISIDYYNMFT	SIFTLCTMSV	DRYIAVCHPV	KALDFRTPRN	AKIINVCNWI
human kappa	ISIDYYNMFT	SIFTLTMMSV	DRYIAVCHPV	KALDFRTPLK	AKIINICIWL

	205	215	225	235	245
human delta	LASGVGVPIM	VMAVTRPRD-	-GAVVCMLQF	PSPSW-YWDT	VTKICVFLFA
human mu	LSSAIGLPVM	FMATTKYRQ-	-GSIDCTLTF	SHPTW-YWEN	LLKICVFIFA
human mu splice	LSSAIGLPVM	FIATTKYRQ-	-GSIDCTLTF	SHPTW-YWEN	LLKICVFIFA
human kappa	LSSSVGISAI	VLGGTKVRED	VDVIECSLQF	PDDDYSWWDL	FMKICVFIFA

	255	265	275	285	295
human delta	FVVPILIITV	CYGLMLLRLR	SVRLLSGSKE	KDRSLRRITR	MVLVVVGAFV
human mu	FIMPVLIITV	CYGLMILRLK	SVRMLSGSKE	KDRNLRRITR	MVLVVVAVFI
human mu splice	FIMPVLIITV	CYGLMILRLK	SVRMLSGSKE	KDRNLRRITR	MVLVVVAVFI
human kappa	FVIPVLIIIV	CYTLMILRLK	SVRLLSGSRE	KDRNLRRITR	LVLVVVAVFV

	305	315	325	335	345
human delta	VCWAPIHIFV	IVWTLVDIDR	RDPLVVAALH	LCIALGYANS	SLNPVLYAFL
human mu	VCWTPIHIYV	IIKALVTIPE	TTFQTVSW-H	FCIALGYTNS	CLNPVLYAFL
human mu splice	VCWTPIHIYV	IIKALVTIPE	TTFQTVSW-H	FCIALGYTNS	CLNPVLYAFL
human kappa	VCWTPIHIFI	LVEALGSTSH	STAALSSY-Y	FCIALGYTNS	SLNPILYAFL

	355	365	375	385	395
human delta	DENFKRCFRQ	LCRKPCGRPD	PSSFSRAREA	TARERVTACT	PSDGPGGGAA
human mu	DENFKRCFRE	FCIPTSSNIE	QQNSTRIRQN	TRDHPSTANT	VDRTNHQLEN
human mu splice	DENFKRCFRE	FCIPTSSNIE	QQNSTRIRQN	TRDHPSTANT	VDRTNHQVRS
human kappa	DENFKRCFRD	FCFPLKMRME	RQSTSRVRNT	VQDPAYLRDI	DGMNKPV*--

	405
human delta	A*--------
human mu	LEAETAPLP*
human mu splice	L*--------
human kappa	----------

Fig. 3.1A.

	5	15	25	35	45
mouse delta	MELVPSARAE	LQSS-PL---	----------	-----VNLSD	AFPSAFPSAG
mouse kappa	ME-SPIQIFR	GDPGPTCSPS	ACLLPNSSSW	FP----NWAE	S-----DSNG
rat mu	MDSSTGPGNT	SDCSDPLAQA	SCS-PAPGSW	LNLSHVDGNQ	SDPCGLNRTG

	55	65	75	85	95
mouse delta	ANASGSPGAR	SAS-SLALAI	AITALYSAVC	AVGLLGNVLV	MFGIVRYTKL
mouse kappa	SVGSEDQQLE	SAHISPAIPV	IITAVYSVVF	VVGLVGNSLV	MFVIIRYTKM
rat mu	LGGNDSLCPQ	TGSPSMVTAI	TIMALYSIVC	VVGLFGNFLV	MYVIVRYTKM

	105	115	125	135	145
mouse delta	KTATNIYIFN	LALADALATS	TLPFQSAKYL	METWPFGELL	CKAVLSIDYY
mouse kappa	KTATNIYIFN	LALADALVTT	TMPFQSAVYL	MNSWPFGDVL	CKIVISIDYY
rat mu	KTATNIYIFN	LALADALATS	TLPFQSVNYL	MGTWPFGTIL	CKIVISIDYY

	155	165	175	185	195
mouse delta	NMFTSIFTLT	MMSVDRYIAV	CHPVKALDFR	TPAKAKLINI	CIWVLASGVG
mouse kappa	NMFTSIFTLT	MMSVDRYIAV	CHPVKALDFR	TPLKAKIINI	CIWLLASSVG
rat mu	NMFTSIFTLC	TMSVDRYIAV	CHPVKALDFR	TPRNAKIVNV	CNWILSSAIG

	205	215	225	235	245
mouse delta	VPIMVMAVTQ	PRDGA--VVC	MLQFPSP--S	WYWDTVTKIC	VFLFAFVVPI
mouse kappa	ISAIVLGGTK	VREDVDVIEC	SLQFPDDEYS	W-WDLFMKIC	VFVFAFVIPV
rat mu	LPVMFMATTK	YRQGS--IDC	TLTFSHP--T	WYWENLLKIC	VFIFAFIMPI

	255	265	275	285	295
mouse delta	LIITVCYGLM	LLRLRSVRLL	SGSKEKDRSL	RRITRMVLVV	VGAFVVCWAP
mouse kappa	LIIIVCYTLM	ILRLKSVRLL	SGSREKDRNL	RRITKLVLVV	VAVFIICWTP
rat mu	LIITVCYGLM	ILRLKSVRML	SGSKEKDRNL	RRITRMVLVV	VAVFIVCWTP

	305	315	325	335	345
mouse delta	IHIFVIVWTL	VDINRRDPLV	VAALHLCIAL	GYANSSLNPV	LYAFLDENFK
mouse kappa	IHIFILVEAL	GSTSHSTA-A	LSSYYFCIAL	GYTNSSLNPV	LYAFLDENFK
rat mu	IHIYVIIKAL	ITI-PETTFQ	TVSWHFCIAL	GYTNSCLNPV	LYAFLDENFK

	355	365	375	385	395
mouse delta	RCFRQLCRTP	CGRQEPGSLR	RPRQATTRER	VTACTPSD--	-------GPG
mouse kappa	RCFRDFCFPI	KMRMERQSTN	RVRN-TVQDP	AS--------	---MRDVGGM
rat mu	RCFREFCIPT	SSTIEQQNST	RVRQNT-REH	PSTANTVDRT	NHQLENLEAE

	405
mouse delta	GGAAA*
mouse kappa	NKPV*-
rat mu	TAPLP*

Fig. 3.1. (part A, opposite page; part B, above) Predicted amino acid sequences of the cloned human (part A) and rodent (part B) opiate receptors.
Amino acid sequences that are shaded refer to identical amino acids among the different receptors. Gaps introduced to generate this alignment are represented by dashes.

between 30-40%.[21] The similarities with somatostatin receptors were particularly interesting since some analogs of somatostatin bind potently to mu opiate receptors[22] and the highly selective mu antagonist CTOP was derived from a somatostatin analog SMS 201-995.[23]

The amino acid sequence of the delta receptor predicts a protein with seven transmembrane spanning regions. There are potential glycosylation sites in the N-terminal region and a predicted disulfide bridge between the second and third extracellular loops. The intracellular loops of the delta receptor are relatively small compared to the adrenergic receptors; this is a common characteristic of peptide receptors. These intracellular regions may be involved in coupling the opiate receptors to G proteins and cellular effector systems.[24]

Following the cloning of the delta receptor, both the kappa and mu receptors were cloned. The mu receptor was cloned using probes against the delta receptor to screen rat brain cDNA libraries.[25-28] The mouse kappa receptor was initially cloned using probes against conserved regions of somatostatin receptors to screen a mouse brain cDNA library.[29] This approach was successful because of the similarities of the amino acid sequences of the somatostatin and opiate receptors.[21] The rodent and human opiate receptors have been cloned and in general there is very little species variation in amino acid sequences.[20,26,30-33]

The predicted amino acid sequences of the three opiate receptors are approximately 65% identical with greatest similarity in the transmembrane spanning regions and intracellular loops and greatest divergence found in the amino- and carboxy-termini and the extracellular loops (see Fig. 3.1). Those regions of amino acid sequence divergence may be critical in causing the functional variations among the receptors.

To test whether the ligand selectivities of the cloned receptors are similar to the endogenously expressed opiate receptors, the cloned receptors have either been stably expressed in different fibroblast cell lines or transiently expressed in Cos cells.[34,35] Ligand binding studies have revealed that the cloned kappa and mu receptors have similar rank order of potencies in binding a large series of opiates as the endogenously expressed receptors.[34] Since the cellular environment of the cloned receptors is different to that of the endogenously expressed receptors, the ligand selectivities of the mu and kappa receptor must be inherent in the receptors and thus strongly dependent on their primary structures.

The pharmacological properties of the cloned delta receptor, on the other hand, did not correlate well with the endogenously expressed receptors.[34] This may be due to the expression of subtypes of delta receptors endogenously, with only one of the subtypes corresponding to the cloned receptor. In fact, the pharmacological properties of the cloned receptor corresponds to a $delta_2$ subtype. NTB, the selective $delta_2$ antagonist, is fifty fold more potent at binding to the cloned delta receptor than the $delta_1$ selective antagonist BNTX. Similarly, deltorphin II, a $delta_2$ selective agonist is 4-5 fold more potent at binding to the cloned receptor than the $delta_1$ selective agonist DPDPE. These findings raise the possibility that either another delta receptor gene exists, that alternative splicing may create a mRNA generating a different receptor protein, or that differential post translational modifications of a common gene product may create the other receptor subtype.

Behavioral studies using antisense against the cloned delta receptor have suggested that distinct mRNAs may encode different delta receptor subtypes.[36,37] Antisense against the delta receptor administered centrally reduced the ability of the $delta_2$ agonist deltorphin II to induce analgesia whereas DPDPE-induced analgesia was less affected. In contrast, administration of the antisense intrathecally blocked both deltorphin II and DPDPE induced analgesia. RNA blotting with delta receptor probes have revealed multiple delta receptor RNA species of high molecular weight, which could encode different proteins.[19,20] However, no definitive evidence exists that different mRNAs encode distinct delta receptors.

Splice variants of the mu opiate receptor have been identified.[38] The variants differ in their C-terminal region. The splice variant is seven amino acids shorter in length and five amino acids in the C-terminal differ in sequence from the unspliced form. The ligand binding properties of the two variants are similar and no clear differences in functional properties have been reported. The gene structure of the mu receptor has been identified and predicts that multiple exons encode the receptor.[39]

The three cloned opiate receptors all couple to adenylyl cyclase and mediate the inhibition of forskolin stimulated cAMP accumulation.[19,20,29,40,41] The kappa receptor has been shown to couple to an N-type Ca^{2+} channel when expressed in PC12 cells.[42] The kappa receptor endogenously expressed in neurons has also been found to couple to this channel and to inhibit neurotransmitter release.[43,44] Both the cloned kappa and mu receptors have been reported to mediate agonist induced increases in K^+ conductance via an inwardly rectifying K^+ channel.[40-45] Studies in oocytes have shown that the receptors couple with the recently cloned inward rectifier GIRK1[47-50] and studies in AtT-20 cells have found that the cloned receptors interact with an endogenously expressed inward rectifier.[51] Previous studies have reported that endogenously expressed mu, delta and kappa receptors couple to an inward rectifier to modulate neuronal excitability and firing rates.[41-46] Thus, both the pharmacological and functional characteristics of the cloned and endogenously expressed opiate receptors are similar.

3.4 STRUCTURE-FUNCTION ANALYSIS OF THE CLONED OPIATE RECEPTORS

The availability of the cloned opiate receptors has provided researchers the opportunity to investigate structural aspects of the receptors involved in the functioning of the receptors. Using site-directed mutagenesis techniques, it has been possible to identify domains of the opiate receptors involved in ligand binding. The first mutagenesis study tested the functional role of a conserved aspartate, at residue 95, in the second transmembrane spanning region of the delta receptor in the binding of ligands (see Fig. 3.2).[52] This conserved aspartate had been reported to be critical for Na^+ to regulate agonist binding to other receptors, including the alpha$_2$-adrenergic[53] and somatostatin[54] receptors. Mutation of this negatively charged amino acid to an asparagine abolished Na^+ regulation of agonist binding to the delta receptor, consistent with the role of this residue as a Na^+ recognition site.[52] However, this amino acid had additional roles in the delta receptor, since the mutant receptor had much lower affinity for selective agonists such as DPDPE but had similar affinity for antagonists as the wild-type receptor. Since the second transmembrane spanning region of the delta receptor has a very similar amino acid sequence to the kappa and mu receptors, it is unlikely that the aspartate 95 is part of the recognition site of the receptor for selective agonists, because one might expect such a recognition site to have a unique amino acid sequence. The aspartate may, however, have a role in establishing the high affinity of the receptor for selective agonists by slowing the rate of dissociation of the agonists from the receptor, through electrostatic interactions. The lack of effect that the mutation of the aspartate to an asparagine has on antagonist binding indicates that agonists and antagonists have different determinants for binding to this receptor.

In addition to the conserved aspartate in the second transmembrane spanning region, the aspartate at residue 128 in the third transmembrane spanning region (see Fig. 3.2) is also involved in agonist binding to the delta receptor (unpublished results of F. Livingston and T. Reisine). Mutation of this conserved residue to an asparagine abolished high affinity agonist binding to the receptor but did not affect antagonist binding. Mutations of this residue in the beta-adrenergic receptor[55] also significantly affected agonist binding. It has been proposed that the negative charge of the aspartate may interact with cationic residues of agonists to facilitate binding through electrostatic interactions. This may occur for the delta receptor also, however, opiate antagonists also have cationic groups needed for high affinity binding and their interaction with the delta receptor was not significantly affected by the removal of the negative charge from the aspartate. Like the aspartate in the second transmembrane spanning region, the aspartate 128 of the delta receptor may also selectively facilitate agonist binding by reducing rates of dissociation of agonists, once agonist is bound to the receptor.

The critical role of the conserved aspartates in the second and third transmembrane spanning regions of the delta receptors are also apparent in the kappa[56] and mu[57] receptors, where mutations of these residues to asparagines greatly diminish agonist binding affinities. Taken together,

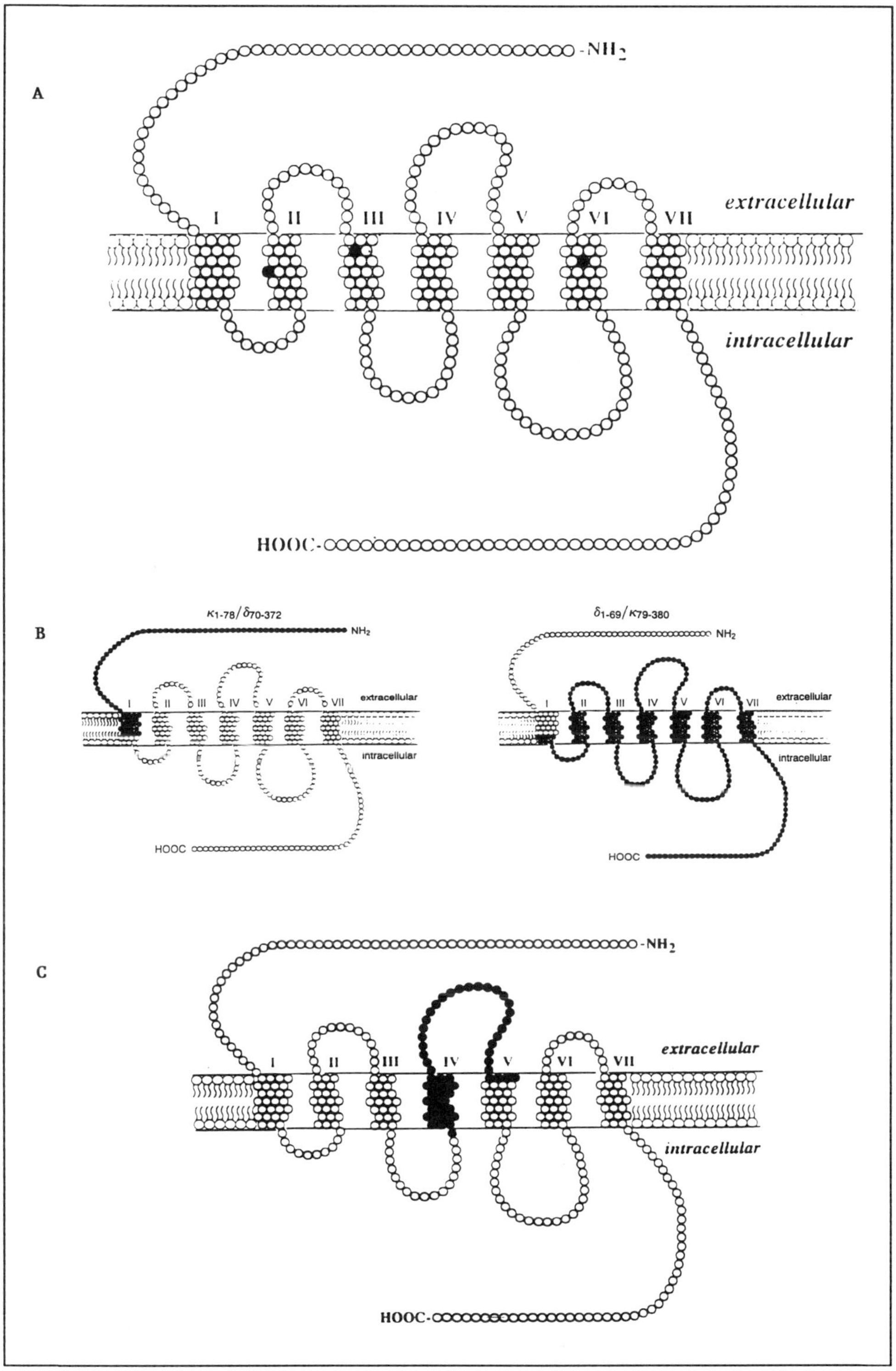

Fig. 3.2. Sequence of mutant opiate receptors. Part A: Schematic of the cloned mouse delta opioid receptor. Dark circles refer to aspartates mutated to asparagine in the second and third transmembrane spanning region and a histidine in transmembrane six mutated to an asparagine. Part B: Chimeric receptors in which regions of the kappa and delta opioid receptors have been exchanged. Dark circles refer to parts of the kappa receptor and open circles refer to the delta receptor. Part C: Chimeric receptor consisting of the kappa receptor (open circles) and a portion of the delta receptor (dark circles).

these results[52,56,57] suggest that these charged residues may have conserved roles in opiate receptors for agonist binding.

Further evidence that agonists and antagonists have different requirements for binding to the opiate receptors was reported in the study of Kong et al[58] in which chimeric receptors consisting of parts of the kappa and delta receptors were generated (see Fig. 3.2B). The N-terminus of the kappa and delta receptor have different amino acid sequences. To test whether the N-terminus of either of these receptors may contribute to ligand binding, the N-termini of each receptor was swapped so that a delta receptor containing the N-terminus of the kappa receptor was generated. This chimera bound delta agonists and antagonists with high affinity similar to their binding to the wild-type delta receptor. Since delta selective ligands do not bind to the kappa receptor, their ability to bind to the chimeric receptor suggests that the N-terminus of the delta receptor is not essential for ligand binding. Similar results have been reported for the mu receptor[41] In contrast, this chimeric receptor bound ^{3}H-naloxone with high affinity and both naloxone and the selective kappa receptor antagonist nor-BNI potently displaced ^{3}H-naloxone binding to the chimera, with affinities in line with their binding to the wild-type kappa receptor. Neither nor-BNI nor naloxone bind potently to the delta receptor. These results suggest that the N-terminus of the kappa receptor conferred onto the delta receptor the ability to bind with high affinity naloxone and the kappa selective antagonist nor-BNI. These findings could be due to the mutations causing unusual conformational changes in the delta receptor such that it was able to bind nor-BNI with abnormally high affinity. This is not likely because the chimera bound delta ligands with affinities similar to their binding to the wild-type delta receptor, suggesting that gross conformational changes in the receptor had not occurred that would affect all ligand binding. The simplest explanation of the results is that N-terminus of the kappa receptor contains recognition sites for antagonists.

In contrast, the kappa selective agonist ^{3}H-U69,593 did not effectively bind to the chimeric receptor and U50,488 did not displace ^{3}H-naloxone binding to the chimera. These findings suggest that agonists and antagonists have different determinants for binding to the kappa receptor.

Further evidence that agonists and antagonists have different domains for binding to the kappa receptors has come from mutagenesis studies designed to identify agonist binding domains of this receptor. Both Xue et al[59] and Wang et al[60] have generated chimeric receptors in which the second extra cellular loop of the kappa receptor was inserted to a corresponding region of the mu receptor. The chimeric receptor expressed higher affinity for Dynorphin A than the wild-type mu receptor, suggesting this region contained recognition sites for this kappa selective peptide. Kong et al[61] generated a different chimeric receptor consisting of the transmembrane 4 and the 2nd extracellular loop of the delta receptor and the remainder of the kappa receptor (see Fig. 3.2C). This receptor bound kappa selective antagonists with high affinity as well as non-selective agonists such as levorphanol. However, peptide and non-peptide kappa selective agonists did not bind to this receptor. The lack of binding of Dynorphin A to this chimera is consistent with the reports of Xue et al[59] and Wang et al[60] and suggest that the second extracellular loop of the kappa receptor contains recognition sites for peptide agonists. The finding that U50,488 and non-peptide kappa selective agonists do not bind to this chimera differs from the results of these other investigators[59,60] However, the mutations involved in the study of Kong et al[61] were unique and encompassed transmembrane 4 as well as the 2nd extracellular loop. Conceivably transmembrane 4 may contain recognition sites for non-peptide kappa selective agonists.

The results of mutagenesis studies on the kappa receptor suggest that agonists and antagonists bind to physically distinct sites within the receptor. Antagonists require the N-terminal region for high affinity binding whereas regions encompassing and surrounding the 2nd extracellular loop are critical for the binding of selective agonists to this receptor. Because the recognition sites do not overlap, competition of agonist and antagonist binding to this receptor must involve steric hindrance or ligand induced conformational changes in the receptor. In fact, in studies on the chimeric receptor consisting of the N-terminus of the kappa receptor and the remainder of the delta receptor, 3H-naloxone binding was potently displaced by delta agonists and antagonists, suggesting that the antagonist binding domains of the kappa receptor were juxtaposed to the delta receptor recognition sites so that competition of ligand binding was possible.

3.5 G PROTEIN COUPLING TO OPIATE RECEPTORS

Opiate receptors couple to adenylyl cyclase and ionic conductance channels via G proteins.[9,62] Many families of G proteins have been identified and cloned. The ones most relevant to the actions of opiates are G_i/G_o, since these G proteins are sensitive to pertussis toxin, and this agent blocks the ability of opiates to inhibit cAMP formation and modulate ionic conductance.

There are three forms of alpha subunits of G_i, that are encoded by distinct genes and that are 85-94% identical in amino acid sequence. There are two forms of G_o that are generated by differential splicing. While opiate receptors have been proposed to interact with all of these G proteins,[10] careful examination reveals that specificity exists in the association of the receptors and the individual alpha subunits.

Different G proteins may link delta receptors to distinct cellular effector systems. Taussig et al[63] have proposed that G_{oa} couples delta receptors in NG-108 cells to Ca^{2+} channels. In contrast, G_{ia1} and/or

G_{ia2} have been proposed to link delta receptors to adenylyl cyclase.[64,65] G proteins may provide the molecular basis for the diversity of action of opiates.

To investigate coupling of the cloned delta receptor with G proteins, the receptors expressed in CHO cells were solubilized and immunoprecipitated with peptide directed antisera against different alpha subunits of G_i and G_o.[66] The receptor was found to associate with G_{ia1}, G_{ia3} and G_{oa}. Upon binding of agonist to the receptor, changes in association of the receptor with the G_i subunits was observed. G_{ia2} associated with the receptor and G_{ia1} dissociated from the receptor. These changes may reflect dynamic events involved in the activation of the delta receptor signal transduction pathway.

While delta receptors associate with G_i and G_o, they are also able to interact with pertussis toxin insensitive G proteins. High affinity agonist binding to the cloned delta receptor is reduced approximately 50% by GTP analogs. However, pertussis toxin treatment only reduced part of that binding, indicating a pertussis toxin insensitive component of the GTP regulation. Peptide directed antisera against G_{za} and G_{qa}, two pertussis toxin-insensitive alpha subunits, were able to immunoprecipitate the cloned delta receptor. These alpha subunits may mediate pertussis toxin insensitive actions of delta agonists. In fact, recent studies by Tsu et al[67] have shown that G_{za} couples the cloned delta receptor to adenylyl cyclase in HEK293 cells and can couple the receptor to phospholipase C in Ltk^- cells. Furthermore, the cloned delta receptor associates with both the $beta_1$ and $beta_2$ subunits of G proteins, and these subunits may also be capable of mediating stimulatory effects of delta agonists on phospholipase C activity.

Kappa receptors, like delta receptors, are capable of coupling to G_{ia} and G_{oa}.[68] In the absence of agonist, the cloned mouse kappa receptor associates with G_{ia1} and G_{ia3}. The binding of agonist to the receptor facilitates G_{ia2} association with the receptor, similar to the delta receptor. Unlike the delta receptor, agonist binding to the kappa receptor does not cause dissociation of G_{ia1} from the receptor. The cloned kappa receptor also associates with G_{za} and G_{qa}, which may mediate pertussis toxin insensitive actions of kappa agonists.

3.6 REGULATION OF OPIATE RECEPTORS

A major limitation of the clinical effectiveness of opiates is tolerance development. Tolerance occurs after repeated administration of opiates such that the initial magnitude of response to a given concentration of opiate diminishes with each drug challenge. In the clinical setting, tolerance is overcome by simply administering higher doses of opiate.[2] However, with time, the high concentrations of opiate can become toxic and life threatening.

The molecular basis of tolerance development to opiates is not well understood.[9,10,62] Biochemical studies have focused on the opiate receptors and their alterations after repeated or continuous opiate treatment. Various studies have reported that repeated opiate treatment to animals or cells expressing the receptors causes opiate receptor desensitization and/or downregulation.

Delta receptors have been shown to desensitize and downregulate following prolonged agonist treatment.[69] The desensitization and downregulation of the cloned delta receptor overlap temporally suggesting that both phenomena may involve common cellular mechanisms or at least involve mechanisms that cannot be easily distinguished.

Few studies have been done on the regulation of kappa receptors since few cell lines endogenously express the receptor and most animal species adaptable to regulation studies have low levels of this receptor (only guinea pig and human have reasonably high enough levels of kappa receptors expressed in their central nervous systems). Studies on the cloned mouse kappa receptor have shown that it can be regulated by continuous agonist treatment.[70] Treatment of the receptor expressed in COS cells for several hours with high concentrations of U50,488 reduced high affinity agonist binding, abolished GTP regulation of agonist binding, and diminished coupling of the receptor to adenylyl cyclase. Under conditions of receptor desensitization, the receptor number was not significantly reduced.

Electrophysiological responses of the cloned kappa receptor also desensitize. Pretreatment of the cloned mouse kappa receptor expressed in AtT-20 cells for 1 hr with U50,488 abolished the ability of the receptor to mediate subsequent stimulation of the inwardly rectifying K^+ current by U50,488. The desensitization was selective because it could be blocked by pretreatment with the antagonist nor-BNI, and no cross-desensitization of somatostatin receptor coupling to the inward rectify was observed, indicating that the desensitization was homologous. Similarly, coupling of the cloned kappa receptor with the cloned inwardly rectifying K^+ channel, GIRKI, in oocytes desensitized following continuous agonist treatment.[48] The desensitization of the kappa receptor may contribute to the molecular basis for tolerance development of kappa agonists seen in vivo. In fact, Attali et al[71] reported that kappa receptor coupling to adenylyl cyclase desensitized in rat spinal cord following chronic etorphine or U50,488 treatment, indicating that kappa receptors involved in mediating analgesic effects of opiates can desensitize.

The desensitization of the kappa receptor may involve the enzyme beta-adrenergic receptor kinase (βARK).[70,72] This enzyme has been shown to phosphorylate agonist bound receptors and may be a primary mediator of homologous desensitization. A βARK dominant negative mutant, which blocks the ability of wild-type βARK to phosphorylate receptors, prevented the rapid desensitization of the cloned kappa receptor.

Tolerance to the biological actions of μ agonists occurs at the cellular level as well as in vivo with animals and humans. Continuous exposure of the cloned rat μ receptor to agonists uncouples the receptor from an inwardly rectifying K^+ channel.[49-51] Prolonged pretreatment of μ receptors expressed in AtT-20 cells with somatostatin results in heterologous desensitization. The ability of μ agonists and somatostatin agonists to stimulate K^+ conductance is lost.[51] The uncoupling of the μ receptor from the K^+ channel was specific, as the desensitization was blocked by naloxone and no change in basal K^+ conductance was observed.

Kovoor et al[50] have also reported that the cloned μ receptor and the cloned K^+ channel, GIRK-1, are coupled when expressed in oocytes. Prolonged agonist treatment causes the desensitization of opiate stimulation of K^+ conductance. The desensitization is heterologous since an endogenous serotonin receptor, that also couples to the K^+ channel, was desensitized. Chen and Yu[49] have reported the desensitization of μ receptor and K^+ channel in oocytes. These results implicate a cAMP-dependent protein kinase and protein kinase C mechanism, although Kovoor et al[50] did not observe similar effects.

Coupling of the μ receptor to adenylyl cyclase is maintained following prolonged agonist exposure. Raynor et al[35] did not find a desensitization of μ receptor responses in COS-7 cells and Wang et al[73] reported that 12 hr treatment of HEK293 cells stably expressing the μ receptor did not affect the ability of the receptor to mediate agonist inhibition of cAMP accumulation. Rat μ receptor inhibition of adenylyl cyclase in stably transfected AtT-20 cells does not desensitize.[51] In preliminary studies on HEK293 cells that stably express the mouse μ receptor, no densensitization was observed following morphine or levorphanol treatment, although a reduction in receptor density occurs.[71] However, treatment of μ-receptor expressing HEK293 cells with etorphine, an extremely potent but nonselective opiate, does result in a desensitization of adenylyl cyclase. These studies suggest that μ receptor coupling to ion channels and adenylyl cyclase is differentially regulated with distinct molecular mechanisms involved in the actions of the μ selective agonists morphine or DAMGO.

The heterologous nature of the desensitization of the mu receptor/K^+ channel coupling may have important implications with regards to the long-term consequences of chronic mu agonist treatments. The inward rectifying K^+ channel is a critical effector system involved in the regulation of neuronal firing activity. Stimulation of conductance through this channel would diminish firing activity. If long-term mu agonist treatment uncouples this K^+ channel from neurotransmitter regulation, then an important brake on the activity of neurons regulated by mu opiates would be lost. As a consequence, the activity of neuronal circuits linked to the cells affected by chronic mu agonist treatment might be altered. It is of interest that short-term (one hour) kappa agonist treatment did not cause heterologous uncoupling of the inward rectifying K^+ channel from the cloned mouse kappa receptor in AtT-20 cells[51], although chronic, 24 hr pretreatment of cultured rat spinal cord neurons with U50,488 did induce heterologous desensitization of alpha$_2$-adrenergic and cholinergic receptor coupling to adenylyl cyclase, as well as uncouple native opiate receptors from adenylyl cyclase.[71] While kappa agonists are known to induce tolerance, they are less likely to cause dependence than mu agonists such as morphine. The different abilities to induce dependence may be related to different long-term cellular effects of mu and kappa agonists, possibly on K^+ channels or other cellular effectors.

Cross-desensitization of the cloned mu receptor with either the cloned serotonin or somatostatin receptors has been observed in oocytes or cell lines. In addition, Harris and Williams[75] reported cross-desensitization of electrophysiological responses of mu and alpha$_2$-adrenergic receptors in locus coeruleus neurons. These authors indicated that the mu desensitization was mostly homologous although there was at least 30% reduced response of noradrenaline to hyperpolarized locus coeruleus neurons that had been pretreated with enkephalin. These findings suggest that depending on the cells and the receptors under study, varying amounts of heterologous regulation can occur after chronic mu agonist treatment. The lack of a pronounced alpha$_2$-adrenergic receptor cross-desensitization that Harris and Williams[75] have reported in morphine tolerant animals may explain the potential usefulness of alpha$_2$-adrenergic agonist treatment of heroin addicts.

3.7 CONCLUSIONS AND FUTURE PROSPECTS

The cloning of the opiate receptors has opened new avenues of research in the opiate field. Using site-directed mutagenesis, it will be possible to identify the ligand binding and G protein coupling domains of these receptors and reveal the molecular mechanisms involved in the regulation of these receptors. Development of specific chimeric or mutant receptors may also aid in the generation of new opiates with few of the undesirable side-effects of available drugs. For example, if dynorphin peptides and synthetic kappa agonists bind to distinct regions of kappa receptors, it may be possible to construct mutant kappa receptors that only bind dynorphin-like peptides. Such mutant receptors could be useful in drug screening for non-peptide dynorphin analogs which might be expected to have analgesic properties, have limited abuse potential and respiratory depressant effects characteristics of morphine, and not induce dysphoria, which is characteristic of spiradoline (the prototypical kappa selective ligand). Furthermore, identification of the ligand binding domains of the opiates may facilitate rational drug design of a new generation of opiates that could be employed not only as analgesics but also in the prevention of graft rejection and treatment of stroke.

Having available the three cloned opioid receptors expressed individually in clonal cell lines will also give researchers a useful model system in which to study the cellular and biochemical basis of tolerance and dependence. Studies are already beginning to show that the cloned mu receptor is regulated differently than the cloned kappa receptor, with the mu receptor desensitization appearing to involve activation of downstream mechanisms, whereas kappa receptor desensitization appears to be largely homologous. This may be reflected in vivo by the greater propensity of mu-selective agonists relative to kappa agonists to lead to the development of dependence. Determining the cellular processes involved in the regulation of the mu receptor may therefore be important in developing strategies to attenuate the harmful side-effects of mu-selective analgesics such as morphine.

Acknowledgments

This work was supported by NIH grant DA08951.

References

1. Reisine T, Pasternak G. Opioid analgesics and antagonists. In: Hardman J et al, eds. The Pharmacological Basis of Therapeutics, New York, Pergamon Press, 1996; 521-555.
2. Pasternak G. Multiple morphine and enkephalin receptors and the relief of pain. JAMA 1984; 259:1362-1367.
3. Pert C, Snyder S. Opiate receptor: Demonstration in nervous tissue. Science 1973; 179:1011-l014.
4. Hiller J, Pearson, Simon E. Distribution of stereospecific binding of the potent narcotic analgesic etorphine in human brain: Predominance in the limbic system. Res Comm Pathol Pharmacol. 1973: 6:1052-l061.
5. Lord J, Waterfield A, Hughes J, Kosterlitz H. Endogenous opioid peptides: Multiple agonists and receptors. Nature 1977; 267:495.
6. Chang K, Cuatrecasas P. Multiple opiate receptors: Enkephalin and morphine bind to receptors of different specificity. J Biol Chem 1979; 254:2610-2618.
7. Chavkin, C., James, I. and Goldstein, A. Dynorphin is a specific endogenous ligand of the kappa opioid receptor. Science 1982; 215:413-415.
8. Simon E. Opioid receptors and endogenous opioid peptides. Medicinal Res Rev 1991; 11:357-374.
9. Loh H, Smith A. Molecular characterization of opioid receptors. Annu Rev Pharmacol 1990; 30:123-170.
10. Herz A. ed. Opioids I, New York: Springer-Verlag, 1993.
11. Goldstein A, Naidu A. Multiple opioid receptors: ligand selectivity profiles and binding site signatures. Mol Pharmacol 1989; 36:265-272.
12. Lutz R, Pfister H. Opioid receptors and their pharmacological profile. J Receptor Res 1992; 12:267-286.
13. Portoghese PS. Selective nonpeptide opioid antagonists. In: Herz A, ed. Opioids I, New York: Springer-Verlag, 1993; 276-296.
14. Portoghese PS, Moe S, Takemori A. A selective $delta_1$ opioid receptor agonist derived from oxymorphone: evidence for separate recognition sites

for $delta_1$ opioid receptor agonists and antagonists. J Med Chem 1993; 36:2572-2574.
15. Portoghese PS, Sultana M, Nagase H, Takemori A. A highly selective $delta_1$ opioid antagonist: 7-benzylidenenaltrexone. Eur J Pharmacol. 1992; 218:195-196.
16. Schiller PW. Development of receptor selective opioid peptide analogs as pharmacological tools and potential drugs. In: Herz A, ed. Opioids I, New York: Springer-Verlag, 1993; 681-710
17. Sofuoglu M, Portoghese PS, Takemori A. Differential antagonism of delta opioid agonists by naltrindole and its benzofuran analog (NTB) in mice: evidence for delta opioid receptor subtypes. J Pharmacol Expt Therap 1991; 257:676-680.
18. Sofuoglu M, Portoghese PS, Takemori A. Cross-tolerance studies in the spinal cord of beta-FNA-treated mice provides further evidence for delta opioid receptor subtypes. Life Sci. 1991; 49:153-156.
19. Evans C, Keith D, Morrison H, Magendzo K, Edwards R. Cloning of a delta opioid receptor by functional expression. Science 1992; 258:1952-1955.
20. Kieffer BL, Befort K, Gaveriaux-Ruff C, Hirth CG. The delta-opioid receptor: isolation of a cDNA by expression cloning and pharmacological characterization. Proc Natl Acad Sci. (USA) 1992; 89:12048-12052.
21. Reisine T, Bell GI. Molecular biology of opioid receptors. Trends in Neurosci 1993; 16:506-510.
22. Maurer R, Gahwiler B, Buescher H, Hill R, Roemer D. Opiate antagonistic properties of an octapeptide somatostatin analog. Proc Natl Acad Sci (USA) 1982 79:4815-4817.
23. Pelton J, Gulya K, Hruby V, Duckles S, Yamamura HI. Conformationally restricted analogs of somatostatin with high mu-opiate receptor specificity. Proc Natl Acad Sci (USA) 1985; 82:236-239.
24. Dolhman H, Caron M, Strader C, Amlaiky N, Lefkowitz R. A family of receptors coupled to guanine nucleotide regulatory proteins. Biochemistry 1988; 27:1813-1817.
25. Chen Y, Mestek A, Liu J, Hurley J, Yu L. Molecular cloning and functional expression of a mu-opioid receptor from rat brain. Mol Pharmacol 1993; 44:8-12.
26. Fukuda K, Kato S, Mori K, Nishi M, Takeshima H. Primary structures and expression from cDNAs of a rat opioid receptor delta-subtype and mu-subtype. FEBS Lett. 1993; 327:311-314.
27. Thompson RC, Mansour A, Akil H, Watson SJ. Cloning and pharmacological characterization of a rat mu opioid receptor. Neuron 1993; 11:903-913.
28. Wang JB, Imai YS, Eppler CM, Gregor P, Spivak CE, Uhl GR. Mu-opiate receptor -cDNA cloning and expression. Proc Natl Acad Sci. (USA) 1993; 90:l0230-l0234.
29. Yasuda K, Raynor K, Kong H, Breder C, Takeda J, Reisine T, Bell GI. Cloning and functional comparison of kappa and delta opioid receptors from mouse brain. Proc Natl Acad Sci (USA) 1993; 90:6736-6740.

30. Mansson E, Bare L, Yang D. Isolation of a human kappa opioid receptor cDNA from placenta. Biochem Biophys Res Comm 1994; 202:1431-1437.
31. Meng F, Xie G-X, Thompson RC, Mansour A, Goldstein A, Watson SJ, Akil H. Cloning and pharmacological characterization of a rat kappa opioid receptor. Proc Natl Acad Sci. (USA) 1993; 90:9954-9958.
32. Minami M, Toya T, Katao Y, Maekawa K, Nakamura S, Onogi T, Kaneko S, Satoh M. Cloning and expression of a cDNA for the rat kappa-opioid receptor. FEBS Lett. 1993; 329:291-295.
33. Knapp R, Malatynska E, Fang L, Li X, Babin E, Nguyen M, Santuro G, Varga E, Hruby V, Roeske W. Identification of a human delta opioid receptor cloning and expression. Life Sci 1994; 54: PL463-PL469.
34. Raynor K, Kong H, Chen Y, Yasuda K, Yu L, Bell GI, Reisine T. Pharmacological characterization of the cloned kappa, delta and mu-opioid receptor. Mol Pharmacol 1994; 45:330-334.
35. Raynor K, Kong H, Mestek A, Bye L, Tian M, Liu L, Reisine, T. Characterization of the cloned human mu opioid receptor. J Pharmacol Expt Therap 1995; 272:423-428.
36. Lai J, Bilsky E, Bernstein R, Rothman R, Pasternak G, Porreca F. Antisense oligodeoxynucleotide to the cloned delta opioid receptor selectively inhibits supraspinal, but not spinal, antinociceptive effects of [D-ALA^2,GLU]Deltorphin. Regulatory Peptides 1994; 54:159-160.
37. Standifer K, Chien CC., Wahlestedt C., Brown G. and Pasternak G. Selective loss of delta opioid analgesia and binding by antisense oligodeoxynucleotide to a delta opioid receptor. Neuron 1994; 12:805-810.
38. Zimprich A, Bacher B, Hollt, V. Cloning and expression of an isoform of the rmu-opioid receptor (rmuOR1B). Regulatory Peptides 1994; 54:347-348.
39. Kaufman D et al. Characterization of murine mu opioid receptor gene. J Biol Chem 1995; 270: 15877-15883.
40. Chen Y, Mestek A, Liu J, Yu L.: Molecular cloning of a rat kappa opioid receptor reveals sequence similarities in the mu and delta opioid receptors. Biochem. J. 1993 295:625-628.
41. Surratt C, Johnson P, Moriwaki A, Seidleck B, Blaschak C, Wang J, Uhl G. Mu opiate receptor: Charged transmembrane domain amino acids are critical for agonist recognition and intrinsic activity. J Biol Chem 1994; 269:20548-20553.
42. Tallent M, Dichter M, Bell GI, Reisine T. The cloned kappa opioid receptors to an N-type Ca^{++} current in undifferentiated PC-12 cells. Neuroscience 1994; 63:1033-l040.
43. Gross R, MacDonald RL. Dynorphin A selectively reduces a large transient (N-type) calcium current of mouse dorsal root ganglion neurons in cell culture. Proc Natl Acad Sci (USA) 1987; 84:5469-5473.
44. Weisskopf MG, Zalutsky RA, Nicoll RA. The opioid peptide dynorphin mediates heterosynaptic depression of hippocampal mossy fibre synapses and modulates long-term potentiation. Nature 1993; 362:423-427.
45. North RA. Opioid actions on membrane ion channels. In: Herz A, ed. Opioids I, New York: Springer-Verlag, 773-796, 1993.

46. Grudt T, William JT. Kappa opioid receptors also increase a potassium conductance. Proc Natl Acad Sci 1993;90:11429-11432.
47. Henry D, Grandy D, Lester H, Davidson N, Chavkin C. Kappa opioid receptors couple to inwardly rectifying K^+ channels when coexpressed by Xenopus oocytes. Mol Pharmacol 1995; 47:551-557.
48. Ma G, Miller R, Kuznetsov A, Philipson L. Kappa opioid receptor activates an inwardly rectifying K^+ channel by a G protein-linked mechanism: Coexpression in Xenopus oocytes. Mol Pharmacol 1995; 47:1035-1040.
49. Chen Y, Yu L. Differential regulation by cAMP-dependent protein kinase and protein kinase C of the mu opioid receptor coupling to G protein activated K^+ channel. J Biol Chem 1994;269:7839-7842.
50. Kovoor A, Henry D, Chavkin C. Agonist induced desensitization of the mu-opioid receptor coupled to potassium channel (GIRK1). J Biol Chem 1995; 270:589-595.
51. Tallent M, Dichter M, Reisine, T. Coupling of the cloned kappa and mu opioid receptor to the inward rectifier potassium current is differentially regulated. Soc Neurosci Abst 1995; 21:532.12.
52. Kong H, Raynor K, Yasuda K, Moe S, Portoghese PS, Bell GI, Reisine T. A single residue, aspartic acid 95, in the delta opioid receptor specifies selective high affinity agonist binding. J Biol Chem 1993; 268:23055-23058.
53. Horstman D, Brandon S, Wilson A, Guyer C, Cragoe E, Limbird L. An aspartate among G protein receptors confers allosteric regulation of alpha$_2$-adrenergic receptors by sodium. J Biol Chem 1990; 265:21590-21595.
54. Kong H, Raynor K, Yasuda K, Bell GI, Reisine T. Mutation of an aspartate at residue 89 in somatostatin receptor subtype 2 prevents Na^+ regulation of agonist binding but does not alter receptor-G protein association. Mol Pharmacol 1993; 44:380-384.
55. Strader C, Sigal I, Register R, Candelore M, Rands E, Dixon R. Identification of residues required for ligand binding to the beta-adrenergic receptor. Proc Natl Acad Sci (USA) 1987; 84:4384-4388.
56. Kong H, Raynor K, Reisine T. Amino acids in the cloned mouse kappa receptor that are necessary for high affinity agonist binding but not antagonist binding. Regulatory Peptides 1994; 54:155-156.
57. Heerding J, Raynor K, Kong H, Yu L, Reisine T. Mutagenesis reveals that agonists and peptide antagonists bind in fundamentally distinct manners to the rat mu receptor than do nonpeptide antagonists. Regulatory Peptides 1994; 54:119-120.
58. Kong H, Raynor K, Yano H, Takeda J, Bell GI, Reisine T. Agonists and antagonists bind to different domains of the cloned kappa opioid receptor. Proc Natl Acad Sci 1994; 91:8042-8046.
59. Xue J, Chen C, Zhu J, Kunapuli S, DeRiel K, Yu L, Liu-Chen L-Y. Differential binding domains of peptide and non-peptide ligands in the cloned rat kappa opioid receptor. J Biol Chem 1994; 269:30195-30199.
60. Wang JB, Johnson P, Wu JM, Wang FW, Uhl G. Human kappa opiate

receptor second extra cellular loop elevates dynorphin's affinity for human mu/kappa chimeras. J Biol Chem 1994; 269:25966-25969.
61. Kong H, Raynor K, Reisine T. Amino acids in the cloned mouse kappa receptor that are necessary for high affinity agonist binding but not antagonist binding. Regulatory Peptides 1994; 54: 155-156.
62. Childers S. Opioid receptor-coupled second messengers. Life Sci. 1991; 48:991-1000.
63. Taussig R, Sanchez S, Rifo M, Gilman A, Belardetti R. Inhibition of the conotoxin-sensitive calcium current by distinct G proteins. Neuron 1992; 8:799-809.
64. Carter B, Medihradsky F. G_{oa} mediates the coupling of the mu opioid receptor to adenylyl cyclase in cloned neural cells and brain Proc Natl Acad Sci 1993; 90:4062-4066.
65. McKenzie R, Milligan G. Delta-opioid receptor-mediated inhibition of adenylyl cyclase is transduced specifically by the guanine nucleotide binding protein G_{i2}. Biochem J 1990; 267:391-398.
66. Law SF, Reisine T. Changes in the association of G protein subunits with the cloned mouse delta opioid receptor upon agonist stimulation. Molecular Biol of the Cell 1994; 5:2737.
67. Tsu R, Chan J, Wong Y. Regulation of multiple effectors by the cloned delta-opioid receptor: Stimulation of phospholipase C and type II adenylyl cyclase. J Neurochem 1995; 64:2700-2707.
68. Law SF, Reisine T.: Association of the cloned delta, kappa and mu opioid receptors with G proteins. Soc Neurosci Abst 1995; 21: 532.13.
69. Law P, McGinn T, Wich M, Erikson L, Evans C, Loh H.: Analysis of delta-opioid receptor activities stably expressed in CHO cell lines: Function of receptor density? J Pharmacol Expt Therap 1994; 271:1686-1694.
70. Raynor K, Kong H, Hines J, Kong G, Benovic J, Yasuda K, Bell GI, Reisine T. Molecular mechanisms of agonist-induced desensitization of the cloned mouse kappa opioid receptor. J Pharmacol Expt Therap 1994; 270:1381-1386.
71. Attali B, Saya D, Vogel Z. Kappa opiate agonists inhibit adenylyl cyclase and produce heterologous desensitization in rat spinal cord. J Neurochem 1989; 52:360-369.
72. Raynor K, Kong H, Heerding J, Hines J, Benovic J, Reisine T. Differential agonist modulation of the cloned opioid receptors reveals distinct cellular mechanisms of receptor regulation. Regulatory Peptides 1994; 54:239-240.
73 Wang Z, Arden J, Segredo V, Sadee W. Agonist induced constitutive activation of the mu opioid receptor by phosphorylation. Reg Peptides 1994; 54:323-333.
74. Blake A, vonZastrow M, Freeman J, Reisine T. Agonist induced regulation of mu and delta opioid receptors. Soc Neurosci Abst 1995; 21: 532.13.
75. Harris G, Williams J. Transient homologous mu-opioid receptor desensitization in rat locus coeruleus neurons. J Neurosci 1991; 11:2574-2581.

CHAPTER 4

Structure, Function and Regulation of the Beta-Adrenergic Receptors

A.D. Strosberg

4.1. INTRODUCTION

The numerous metabolic and neuroendocrine effects of the catecholamines such as adrenaline and noradrenaline are mediated by a family of membrane-bound proteins known as the adrenaline, adrenergic receptors ("AR"), or "adrenoceptors." Natural or synthetic catecholamines or analogs activate various cellular transduction mechanisms by binding to these receptors, thus leading to activation of GTP binding regulatory "G" proteins which in turn modulate (activate or inhibit) effectors such as adenylyl cyclase or phospholipase C. The nine known adrenergic receptors $\alpha_{1AD,B,C}$, $\alpha_{2A,2B,2C}$ and β_1,β_2,β_3 constitute only a small branch of the much larger family of receptors coupled to G-proteins (for a recent review on adrenergic receptors see Strosberg[1]). This family comprises receptors for a variety of ligands including biogenic monoamines, neuropeptides, peptidic and non peptidic hormones as well as for sensory stimuli such as light, gustatory and olfactory substances. Despite their apparent diversity of function, these proteins share important structural properties: they are all composed of a single polypeptide chain with seven hydrophobic regions likely to constitute seven transmembrane domains spanning the lipid bilayer, and they are all coupled to a GTP binding protein, hence our "R_7G" abbreviation for the family members. This chapter is entirely devoted to the structure, function and regulation of the three β-adrenergic receptors, which constitute examples of the similarities and differences one may find between three subtypes which bind the same natural ligands and modulate the same effector.

Recombinant Cell Surface Receptors: Focal Point for Therapeutic Intervention, edited by Michael J. Browne.

4.2. HISTORY

Ahlquist was the first to propose that the effects of adrenaline and noradrenaline could be ascribed to binding to two different classes of receptors: α and β.[2] Lands further refined this pharmacologic classification by distinguishing α_1 and α_2, and β_1 and β_2 subtypes.[3] While α1 receptors are coupled to phospholipase C modulated Ca^{2+} fluxes, α2 receptors regulate inhibition of adenylyl cyclase and β receptors also modulate activation of this enzyme. All three classes also act upon other biological signals including K^+ and possibly Na^+ fluxes (for a more complete discussion see refs. 1, 4, 5). The existence of several additional subtypes of adrenergic receptors has been uncovered by utilizing as probes the genes or cDNA of the known subtypes to clone homologous genes subsequently shown to encode receptors whose existence had earlier sometimes been suspected because of "atypical" pharmacologic effects. This was namely the case for the β_3AR, the subtype which was first cloned from a human genomic library in 1989.[6]

4.3. PHARMACOLOGY

The pharmacology of and βAR was, for a long time, restricted to that of the β_1 subtype, the predominant βAR in heart and of the β_2 subtype, the major and βAR in lungs, trachea and uterus. These subtypes are the targets of major classes of drugs commonly used in human medicine: the β blockers which are the β_1 antagonists used for treatment of heart disease and β_2 agonists used for the treatment of asthma (see Table 4.1). A wide variety of compounds have been developed to address the needs of selectivity, potency and in vivo stability. Numerous chapters and books have been developed to describe the properties of the well-studied and β_1 and β_2AR, and I will only discuss their characteristics in sofaras they relate to the properties of the newly discovered β_3AR. The evaluation of so called "side-effects" observed in several tissues led to the suggestion of the existence of one or more "atypical βAR" present in fat tissues and also possibly in smooth and skeletal muscle.[7,8] The pharmacologic properties of the recombinant β_3AR account for most of the atypical effects described in adipocytes or fat tissues. β_3AR agonists (see Table 4.1) have thus been shown to promote lipolysis and thermogenesis in rodent and human tissues and adipocytes, even in the presence of potent β_1/β_2 antagonists. The lack of a selective radiolabeled antagonist continues, however, to hamper a clear demonstration of the equivalence of the β_3 and the atypical receptors in a variety of other tissues; this is the case for smooth-muscle from the gut where β_3 agonists induce relaxation.

Undoubtedly, agonists for the β_3 subtype expressed in fat tissues may constitute a major new class of lipolytic drugs for the treatment of obesity and thus diabetes. Whether these will lead to pharmaceutical developments as important as those of β_1 and β_2 ligands remains uncertain, however, especially considering the low concentrations of

Table 4.1. The human β adrenergic receptors

	β1	β2	β3
Pharmacology			
Potent agonist	Iso > bucindolol noradrenaline	Clenbuterol, adre ≥ Iso	Isoproterenol > noradre, bucindolol[b]
Selective agonist	Xamoterol	Procaterol	CL316,243
(Selective) antagonist	CGP20712A	ICI 118551	Bupranolol[b]
Mechanism of Action			
	↑cAMP	↑cAMP	↑cAMP
Gene			
Chromosome	10	5	8
Existence of introns	no	no	yes(1)
Protein Structure			
Length	477	413	408
Glycosylation (N-terminal)	2	2	2
Phosphorylation sites	PKA, βARK	PKA, βARK	0
Prototypic Tissue	heart	lung	fat

adre: adrenaline; noradre: noradrenaline; AC: adenylyl cyclase; IP3: Phosphatidyl inositol triphosphate; PKA: protein kinase A; βARK: β adrenergic receptor kinase (sites predicted on the basis of the consensus target sequences). [b]: Bupranolol is the best β3 antagonist identified so far, but it is not selective for this subtype.

β_3AR in target tissues. The exponential growth of the number of articles[1,7] and patent applications concerning β_3 ligands and their properties would suggest that this is certainly a growing area.

4.4. STRUCTURE OF THE βAR

4.4.1 Affinity Purification

The development in the late 1960s of radioligands and detergents to study the βAR opened the way to the purification of the receptors. In 1977, Vauquelin et al[9] were the first to report an affinity chromatography gel made of alprenolol, a βAR antagonist, coupled to agarose. Turkey erythrocyte membranes, solubilized using digitonin, a mild non-ionic detergent, were absorbed on the column, and adenylyl cyclase was washed off the gel together with over 90% of the other membrane proteins. The turkey erythrocyte equivalent of mammalian βAR was eluted off the column using isoproterenol, a synthetic agonist.

This first successful affinity purification of βAR was further developed by Vauquelin et al[10] and Caron et al.[11]

4.4.2 Cloning and Sequencing of the Gene and cDNA

Cloning and sequencing of the receptor genes and cDNA was based on partial sequencing of tryptic peptides of the purified receptor proteins.[12,13] Partial amino acid sequencing of a few tryptic peptides led to the synthesis of oligonucleotides which were used as probes to clone the corresponding hamster β_2 cDNA,[12] turkey β_1-like cDNA[13] and the human platelet β_{2A} cDNA.[14] Hydropathy plots of the predicted amino acid sequences revealed the presence of the seven hydrophobic segments, previously identified as transmembrane domains in bacteriorhodopsin. Despite the lack of amino acid sequence homology, the similarity with the bacterial light receptor led to a series of fruitful studies which basically sustained the hypothesis that all seven hydrophobic segments contribute to form a ligand binding pocket, as was indeed demonstrated for bacteriorhodopsin by high resolution electron cryoscopy[15] and by biochemical[16] and immunologic techniques[17] for the β-adrenergic receptors. We present in Figure 4.1A, the membrane topography of a typical G-protein coupled mammalian receptor, the human β_2 receptor, the gene of which was cloned by Emorine et al[18] and Kobilka et al[19] based on its homology with hamster cDNA.[12] The human protein is composed of a single polypeptide chain of 413 amino acid residues with an extracellular N-terminus containing two consensus sites for N-linked glycosylation, seven transmembrane segments of 21 to 28 residues, three extra- and three intracellular loops and an intracellular C terminus containing two sites for phosphorylation by protein kinase A, as well as several sites for phosphorylation by the β-adrenergic receptor kinase.

4.4.2.1 Cloning of the β_1AR gene

The successful cloning by cross-hybridization of genes homologous to those encoding the βAR and muscarinic acetylcholine receptors suggested that a similar approach would soon lead to the isolation of gene or cDNA encoding the β_1AR. This approach, however, led to the cloning of a gene, G21 which two years later was shown to encode the 5HT-1A serotonine receptor. To isolate the β_1AR gene we used the turkey erythrocyte β_1 like gene, whereas Frielle et al[20] finally succeeded in cloning this gene by using G21 as a new probe.

4.4.2.2 Cloning of the β_3AR gene

The turkey erythrocyte, β_1AR gene probe also revealed the existence of additional hybridizing DNA fragments from which the unexpected human β_3AR gene was isolated.[21] This gene was initially thought to be intronless, as were the β_1 and β_2AR genes. Several years later however, β_3 genes were shown actually to contain respectively one (man)[22]

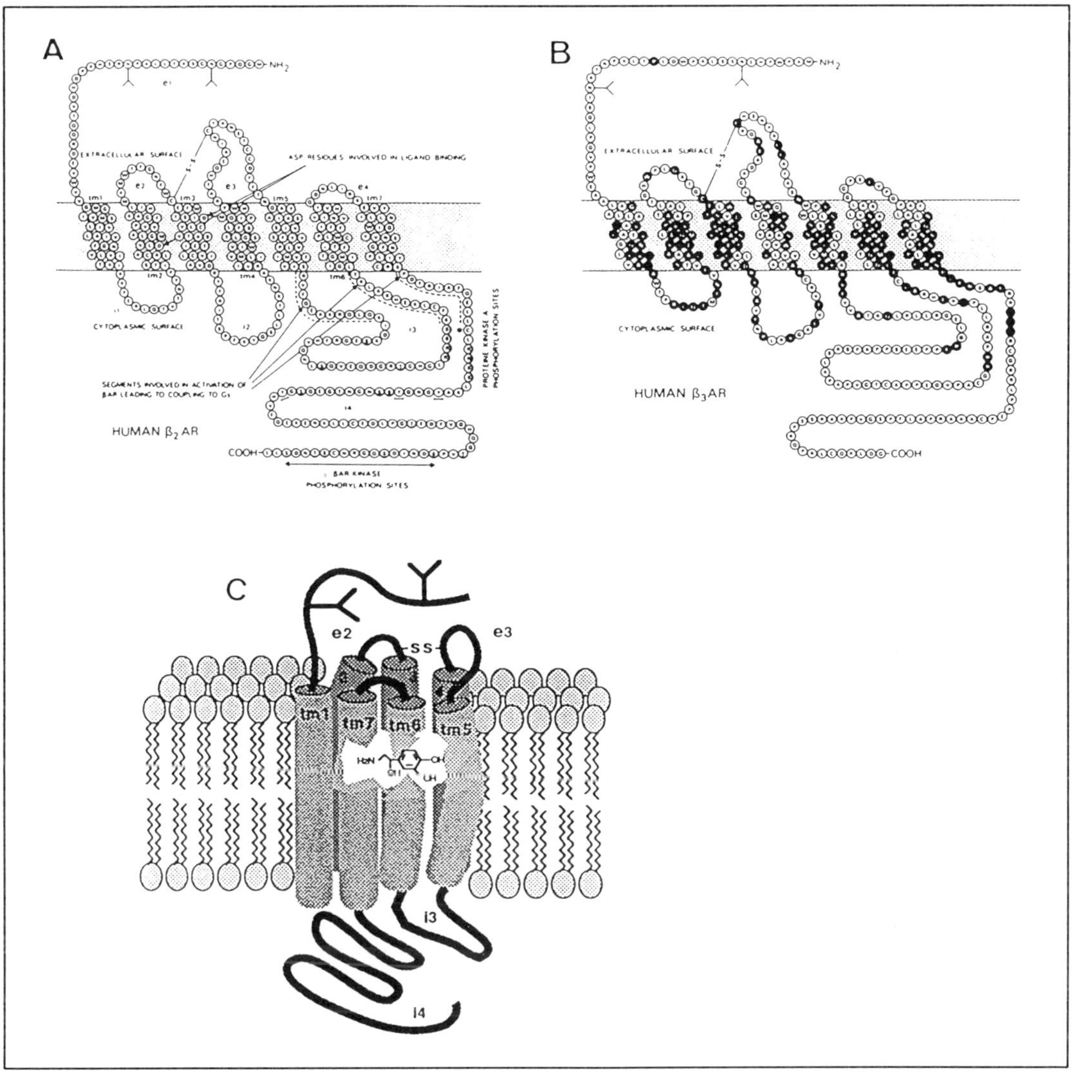

Fig. 4.1. Topological models for β2 and β3-adrenergic receptors. A: Primary structure of the human β_2-, B: β_3-adrenergic receptors and C: Proposed membrane topology of βAR. (A) the sequences are represented in the one-letter code for amino acids. The single polypeptide chain is arranged according to the model for rhodopsin. The disulfide bond, essential for activity linking Cys^{106} to Cys^{184} is represented by -S-S-. The two N-glycosylation sites in the amino-terminal portion of the protein are indicated by Y, i = intracellular loop, e = extracellular loop. The palmitoylated Cys^{341} residue in the N-terminus of the i4 loop is indicated by an asterisk. Potential Ser and Thr phosphorylation sites are underlined. The three Tyr residues found in the i4 of β_2-, but not β_1- or β_3AR are indicated by (-) signs. (B) The residues that are conserved among the β_1, β_2 and β_3 subtypes are indicated in black circles. (C) Model for membrane organization of βARs. The seven transmembrane domains are shown to interact with each other and thus to form the ligand-binding region which is buried in the lipidic bilayer.[65] Recent evidence suggests that in peptide receptors, extra cellular loops may also contribute to the receptor-ligand interaction. In βAR, the disulfide loop (-S-S-) between e_2 and e_3 is certainly essential for activity.Reprinted with permission from Neuroprotocols 1994; 4:32-40.

and two (rodents)[23] introns and thus additional exons encoding respectively, 6 residues (in man) and 12 residues (in rodents). The role of the β_3 introns and exons has not yet been elucidated.

4.5 EXPRESSION OF THE CLONED βAR RECEPTOR GENES

One of the major goals of cloning the βAR genes was to express them in model systems in order to produce enough recombinant receptor protein in order to perform detailed structure-function relationship studies. For this purpose, several expression systems were used:

4.5.1 Mammalian Expression

The foremost and today still the most useful system to express βAR genes are the Chinese Hamster Ovary cell (CHO), the Chinese Hamster Fibroblast cell line (CHW), and murine fibroblast cells (L cells). In these cells the expression levels may be varied almost at will, thus yielding relatively low (20 to 50,000) numbers of receptors per cell compatible with "normal" coupling to the appropriate G protein and adenylyl cyclase effector. These same cells may also be used to obtain very high numbers of receptors (up to a 10^6 or more per cell), very useful for binding studies, but where coupling is distorted (some authors say "promiscuous"...).

The major advantage of the CHO cells is that they produce homogeneous populations of βAR proteins which appear to be fully functional and glycosylated.

4.5.2 Expression in Baculovirus-Infected Insect Cells

Sf9 cells may be infected with recombinant baculovirus containing genes encoding the β_1 or β_2AR[24,25] and very high levels of corresponding receptor protein may thus be produced. Unfortunately, most of these receptors (up to 90%) are not folded appropriately, and are only partially glycosylated. To obtain receptors suitable for functional studies, affinity purification on ligand-containing agarose is necessary[26] thus somewhat reducing the interest of this expression system.

4.5.3 Expression in Microorganisms

The fact that *Halobacterium* functionally express a R_7G protein (bacteriorhodopsin) prompted attempts to use *E. coli* bacteria for the expression of human βAR.[27] This proved to be a very fruitful approach, yielding fully functional β_1 and β_2AR, despite their lack of glycosylation.[27,28] These bacterially expressed receptors were shown to display all the pharmacologic properties of their mammalian cell-produced counterparts.[29] Mutant[30] and chimeric receptors[31] could be characterized in these cells which also provide an easy-to-automate screening system for the pharmaceutical industry now routinely used to express not only the βAR but also a number of other R_7G proteins, such as the serotonin 5HT-lA[32] and the neurotensin receptors.[33,34] The bacterially produced βAR can be coupled to G_s[35] confirming its integrity. King et al[36] also reported the expression of the $β_2$AR in yeast where it was artificially coupled to the G protein growth arrest response mediated by the α and a mating factors. Although quite astute, this model could apparently never be reproduced by any other groups attempting to develop similar genetic constructs.

4.6. POLYMORPHISM OF βAR AND CORRELATION WITH DISEASE

4.6.1 $β_2$AR VARIANTS

A number of polymorphic forms of the $β_2$AR have been identified in the general population including three with one or two missense mutations encoding residues in the N-terminal region of the receptor: $Arg^{16} \rightarrow Gly$, $Gln^{27} \rightarrow Glu$ and one form displaying both substitutions at the same time ($Arg^{16}Gly + Gln^{27}Glu$).[37] As expected from their location of the changes, the mutated receptors when expressed in transfected CHO cells, displayed normal agonist binding and functional coupling to G_s followed by stimulation of adenylyl cyclase. In contrast, the degree of agonist-promoted downregulation of receptor expression appeared to be increased for the $Arg^{16} \rightarrow Gly$ and the double mutant while the Gln^{27} to Glu appeared to become resistant to down regulation of $β_2$AR.[37] Another naturally occurring variant occurs in tm4 where Ile substitutes for Thr at position 164 ($Thr^{164} \rightarrow Ile$)[38] thus adjacent to Ser^{165} which had earlier been predicted to interact with the β-carbon hydroxyl group of catecholamines. The mutated receptor, when expressed in CHO cells, indeed displayed a two-fold lower binding affinity for adrenaline and isoproterenol but not for dobutamine or dopamine which lack hydroxyl groups on their β-carbons.

4.6.2 $β_3$AR VARIANT

A single point mutation at position 64, at the end of tm1 which substitutes a Trp of the $β_3$AR by an Arg residue was found with an allelic frequency of 8 to 10% of the general population and up to 30% of North American Pima Indians.[39] This isolated population is characterized by an unusual high incidence of obesity and diabetes, which appears to be hereditary. While no effect of the mutation has so far been detected in non obese individuals, the influence appears rather striking in obese patients who display an increased dynamic capacity to gain weight[40] and to develop subsequent diabetes.[41]

The mutant receptor, when expressed in CHO or L cells, appears to bind the ligand with a K_D very close to that of the unchanged $β_3$, but coupling appears to be reduced. It is interesting to note that the mutation introduces in man the Arg residue which is found in an equivalent position in the $β_3$AR of all other species studied so far.

While obesity and diabetes clearly are polygenic diseases, the Trp64→Arg may reveal a risk factor which should be detected at an early time. A PCR based diagnostic test may constitute an appropriate early warning system to be offered to all populations at risk.

4.7. TRANSGENIC MICE EXPRESSING HUMAN βAR GENES

4.7.1 Human β_1AR Transgenic Mice

The gene encoding the human β_1AR was expressed in transgenic mice under the control of a 0.56 kb proximal promoter of the human atrial natriuretic factor.[42] The human receptor was found to be over expressed specifically in the left and right atria of the mice. The level of β_1AR was 5 to 10 fold higher in transgenic animals compared to basal levels of murine β_1AR in control mice. The human β_1AR was also able to couple to murine G_s. Electrocardiographic measurements revealed supraventricular premature beats in 6/14 transgenic mice compared to 2/16 control mice, suggesting pathophysiologic effects of over activation of atrial β_1AR responsive signalling system. Treatment of the animals with β_1AR agonists or antagonists, respectively, revealed a decreased response in terms of atrial contractility and heart rate variability.[43]

4.7.2 Human β_2AR Transgenic Mice

The gene encoding the human β_2AR was expressed in transgenic mice under the control of the murine alpha myosin heavy chain promoter.[44] Cardiac overexpression was quantified by ligand binding assays and was 55 to 195 times the amount of βAR in control tissue. This resulted in increased basal myocardial adenylyl cyclase activity, enhanced atrial contractility and increased left ventricular function in vivo; the basal levels of these activities in the β_2AR transgenic animals were equal to the maximal isoproterenol-stimulated levels in control mice. An antagonist with negative efficacy ("inverse agonist") such as ICI-118551 markedly reduced the increased basal responses in the transgenic animals.

The studies of the β_1 and β_2AR transgenic mice illustrate the use of such animals for studying the functional consequences of altered myocardial gene expression in vivo, especially in light of the reported alterations of the βAR system in chronic heart failure.

4.8. LIGAND BINDING AND SIGNAL TRANSMISSION IN ADRENERGIC RECEPTORS

4.8.1 The Ligand Binding Pocket

The β-adrenergic ligand binding site is formed by the seven membrane spanning domains and may be represented as seen in Figure 4.1C: removal of most of the amino or carboxy-terminal residues by proteolysis[17,45] or by deletion mutations of β_2AR[12] or β_3AR[45] has little or no effect on the binding of ligands to the β-adrenergic receptor. The two half cysteines forming the disulfide bond linking extracellular domains e2 and e3 appear to be essential, however.[10,12,16,47] Fluorescence quenching analysis indicated that the β_2 antagonist carazolol is in fact buried at least 10.9 Å deep into the hydrophobic core of the receptor.[48]

Photoaffinity labeling and site-directed mutagenesis studies have helped in the identification of receptor residues belonging to regions in close proximity to the ligand. Two types of residues have thus been identified: those associated with agonist binding and those involved in G protein activation. A number of such residues have been represented in Figure 4.2, which shows key interactions between the β_2-adrenergic receptor and the agonist noradrenaline.

4.8.2 Ligand Binding Residues

The most important residue is undoubtedly Asp^{113} in tm3 which is conserved in all adrenergic receptors, indeed in all monoamine receptors analyzed so far. Its carboxylate group is believed to act as a counter-ion for the amino group present in the ligand: substitution of this aspartate by any other residue except glutamate abolishes binding of monoamines.[12] When a β_2AR mutant was generated in which glutamate replaces aspartate, the β_2 antagonists pindolol and oxprenolol were recognized as partial agonists.[49] Position 113 is clearly crucial for receptor function: when the aspartate is changed to serine, catechol esters, which may form a hydrogen bond with the hydroxyl group of this residues, become able to act as agonists towards the mutant receptor.

The role of hydrogen bonds is also important in other positions: site-directed mutagenesis suggested that Ser^{205} (tm5) forms a hydrogen bond with the metahydroxyl group of the catechol ring and that Ser^{208} (also in tm5) forms a hydrogen bond with the para-hydroxyl group 49.

Mutation of several other residues in different transmembrane domains also affects ligand binding,[12,49,50] including Ser^{165} (tm4), and Phe^{290} (tm6), and the four cysteine residues Cys^{106}, Cys^{184}, Cys^{190} and Cys^{191} mentioned before. While most of these observations were performed on the β_2AR, initial results on β_3AR broadly confirm these results.[46]

4.8.3 Signal Tranduction

While Asp^{113}, Ser^{204} and Ser^{207}, which exist in homologous positions in all adrenergic receptors, appear essential for ligand binding, other residues undoubtedly participate both in binding and in signal transmission, leading to activation of the α subunit of the G protein coupled to the receptor. This is the case for Asp^{79} (tm2): substitution of this residue in the β_2 receptor results in severe loss in affinity for agonists, while antagonist binding remains essentially unchanged.[30,51,52] Three other residues in the β_2AR appear to play a major role in signal transmission. Substitution of Asp^{130} by Asn results in a receptor with high affinity agonist binding that is uncoupled from adenylyl cylase.[53] Two other crucial residues are Tyr^{316} and Asn^{312} (both in tm7). A multistep dynamic model has been proposed[1] in which the agonist induced interaction of Asp^{79} with Tyr^{316} may constitute the first step towards activation of the G protein. Antagonist binding would promote formation of a hydrogen bond between Tyr^{316} and Asn^{312} and thus prevent G_s activation.

4.8.4 Determinants of Subtype-Selective Binding/Signal Transduction

Results from α-β and β_1-β_2 chimeric receptor studies have confirmed that determinants of subtype selectivity are found on several if not all of the seven transmembrane domains. Replacement of the entire tm5-i3-tm6 (i3 = intracellular loop) region of the α_{2A} AR by the corresponding region of the β_2AR resulted in a chimeric receptor capable of stimulating adenylyl cyclase in response to α_2 agonists, albeit with reduced efficiency. A similar chimeric $\alpha 1/\beta_2$AR was able to stimulate both phospholipase C (the α1effector) and adenylyl cyclase.[54]

The seventh domain appears to be important in determining differences in antagonist binding specificity between the β_2- and β_2-adrenergic receptors. Substitution of the Phe^{412} by Asn in the human platelet β_2AR, for example, led to the loss of binding of the β_2 antagonist yohimbine and the acquisition of high affinity for the β_1/β_2 antagonists alprenolol, propranolol and pindolol but not sotalol.[55] Studies of β_1/β_2 chimeric receptors confirmed that all the tm domains appear to contribute residues forming the ligand binding site.[12,20] While these authors suggested a progressive change in relative potency of the β_1 selective antagonist betaxolol and the β_2-selective antagonist ICI-118551 when domains of β_1 were replaced by domains of β_2, a more detailed analysis of an extensive number of chimeric β_1-β_2 receptors functionally expressed in *E. coli* [27,28] demonstrated that in fact each of 11 selective ligands appears to define its own ligand binding subsite.

4.8.5 Site of Interaction with G-Proteins

Residues composing the amino- and carboxy-terminal segments of the third intracellular loop (i3) appear to constitute the main site of interaction with G proteins. This was demonstrated by studying the effects of deletion and homologous replacements by sequences from other receptors. In one such study, Wong et al[17] exchanged a 12 amino acid sequence in the N terminus of i3 of the muscarinic M1 receptor with the homologous stretch from the turkey βAR. Binding of the muscarinic agonist acetylcholine to the M1-β chimera led to activation of both phospholipase C (the M1 effector) and adenylyl cyclase (the β effector). Replacement of the i2 M1 sequence by the turkey βAR sequence in the chimera conserved stimulation of cyclase but reduced activation of phospholipase C. In fact, a single peptide corresponding to the C terminus of i3 of the human β_2AR is sufficient to activate the G_s protein. When this peptide is phosphorylated, as is the case of the whole β_2AR during the process of desensitization, the peptide loses its ability to stimulate G_s but displays an increased ability to stimulate G_i. Substitution of just three residues from the carboxy-terminus of i3 of the α1B_2AR by the homologous β_2AR amino-acids ($Arg^{288} \rightarrow$ Lys, $Lys^{290} \rightarrow$ His and $Ala^{293} \rightarrow$ Leu) resulted in an increase in both the binding affinity of noradrenaline and its potency to stimulate phospholipase C-mediated phospho-inositol turnover by 2 to 3 orders of magnitude and rendered the receptor constitutively active in the absence of agonist-induced activation.[56] Reciprocal mutation in the β_2AR also led to a constitutively activated receptor.[57] The comparison of the three βAR (Fig. 4.1) reveals that most of the residues believed to be responsible for G-protein interaction, including those in i_4 (intracellular loop 4) proximal to the membrane up to Cys^{341}, are well conserved, in line with the finding that all three subtypes interact with the same αs subunit of G_s protein.

4.9 REGULATION OF SUBTYPE EXPRESSION

The coexistence, in the same cells, of several subtypes of receptors which may bind the same natural agonists, albeit with varying affinities, suggests an important role for regulatory mechanisms acting at the level of the gene or the protein.

4.9.1 Developmental Regulation

Genes encoding the various adrenergic receptor subtypes may each respond to different signals during ontogenesis: the majority of β_3 specific mRNA is thus detected in brown adipose tissue which, in mammals other than rodents, is found mainly in newborns or in rare pathological situations such as pheochromocytoma. Isolated brown or white adipocytes found throughout the life-span of human adults do, however, express this β_3AR at the same time as they express β_1 and β_2AR. In rodents, the β_3 subtype is the predominant subtype expressed in brown adipose tissue. In murine 3T3-F44-2A fibroblasts and in human PAZ-6 immortalized PAZ-6 pre-adipocytes[58] only β_1 and β_2 are detected, but when induced to differentiate into adipocytes, these cells start to express β_3AR; in mouse the β_2 actually becomes barely detectable, whereas in man all three subtypes are detected.[58]

4.9.2 Pharmacologic Regulation

In the same murine 3T3-F44-2A adipocyte-like cells, the expression of the β_2AR may be considerably upregulated, and that of the β_1 and β_3AR almost completely suppressed by treatment with the glucocorticoid dexamethasone. This up-regulation of β_2AR, previously described in transfected cells,[18,59] may be explained by the existence in the 5' flanking region of the β_2AR of several glucocorticoid responsive elements (GRE), consensus sequences which are potential sites of interaction with the glucocorticoid receptor. Although analogous sequences may also be recognized in the β_3 5' region, they are situated close to AP-1 binding sites: in other genes, such proximity results in negative regulation by dexamethasone. Cyclic AMP responsive elements (CRE) have also been identified in the 5' flanking region of βAR. In β_2, at least one CRE may be recognized by a specific binding protein ("CRE-BP") leading to increased transcription.[59] In human β_3, three CRE seem effectively to regulate agonist-induced increased transcription of the receptor gene by cAMP.[59]

4.9.3 Regulation by Post-Synthetic Modifications

Phosphorylation by protein kinase A

All adrenergic receptor subtypes, except the β_3AR, contain at least one and sometimes two consensus target sites for phosphorylation by protein kinase A (PKA). These Arg/Lys-Arg-X-(X)-Ser/Thr sequences all occur in the i3 or carboxy-terminal domains, close to the sites of G-protein interaction (Fig. 4.2). Absence of these sites, as in β_3AR [21] or their removal by mutagenesis,[61] almost completely prevents desensitization by low (nanomolar) concentrations of agonist. In a β_2/β_3 AR chimera the reintroduction of PKA and βARK phosphorylation sites partially restored rapid agonist promoted desensitization. In the β_1 and β_2AR, PKA phosphorylation promoted by agonist leads to downregulation of receptor mRNA levels.

4.9.4 Phosphorylation by β–Adrenergic Receptor Kinase

Phosphorylation of agonist-occupied receptors may also occur in mutant β_2AR lacking the PKA sites[61] or in the presence of PKA inhibitors[62] leading again to receptor desensitization, but only in relatively high agonist concentrations (micromolar). This type of phosphorylation is caused by an enzyme which is functionally related to rhodopsin kinase and has been named β-adrenergic receptor kinase (βARK) because it was initially thought to act only on agonist-loaded βAR,[63] although it has now been shown to phosphorylate several other types of G protein-coupled receptors including β_{2A}, muscarinic M2 and even rhodopsin. In contrast to PKA, βARK action does not directly interfere with activation of G_s. However, rhodopsin kinase mediates desensitization of rhodopsin by causing binding of arrestin to the phosphorylated protein, thus disrupting the interaction with transducin, the G protein involved in light adaptation. By analogy, Lohse et al[64] identified a β-arrestin which appeared to interfere with G_s activation of adenylyl cyclase. The Ser/Thr target sites for βARK are most likely located in the carboxy-terminal region of the β_2AR: replacement or deletion of all of the 11 Ser and Thr residues closest to the C-terminus resulted

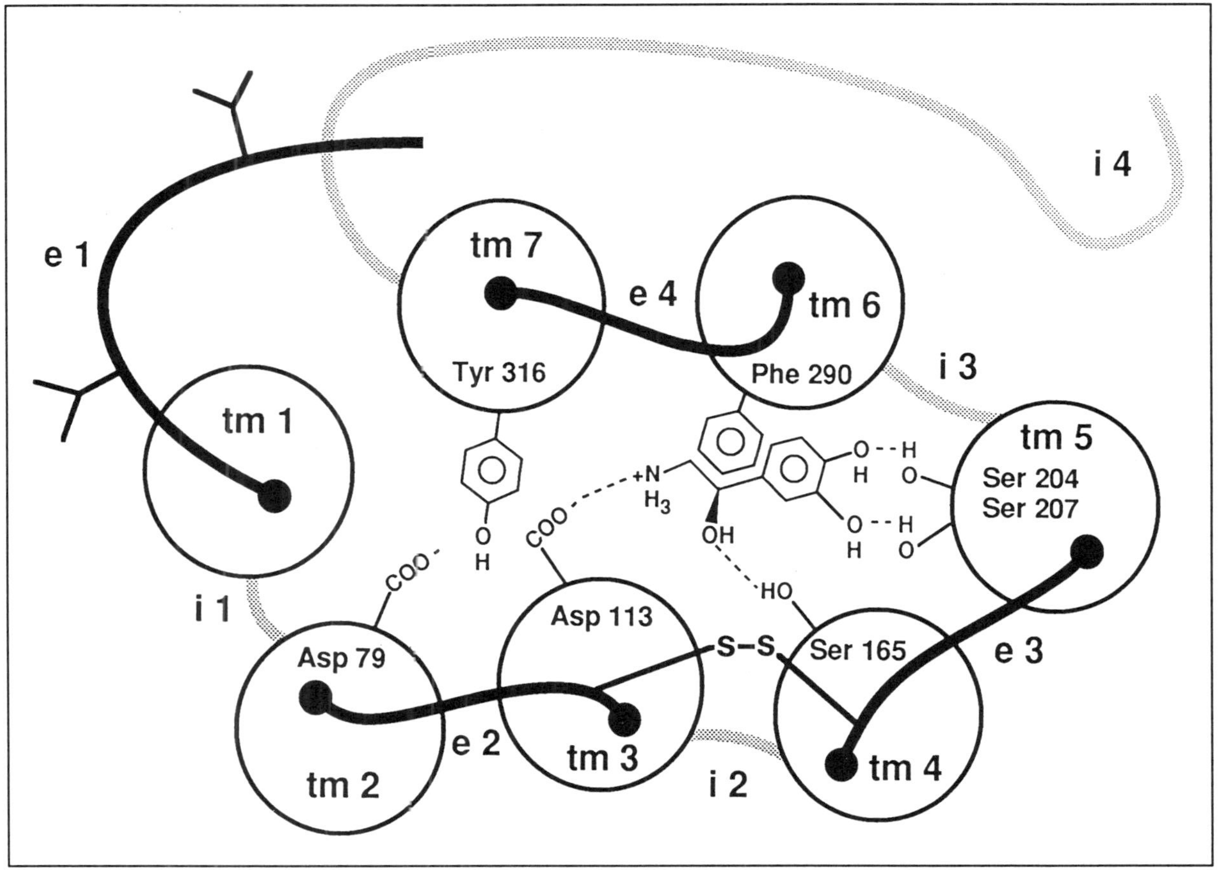

Fig. 4.2. A schematic view of the chemical interactions of noradrenaline with various residues of the β_2-adrenergic receptor binding site. A composite image of the β_2AR ligand-binding region. Proposed interactions in the ligand-binding area of the βAR viewed from the outside of the cell. All seven tm domains are essential for ligand binding. The ligand noradrenaline is shown surrounded by several of the amino acid side chains that are speculated to be involved in agonist binding. These are Asp^{113} in tm3, Ser^{165} in tm4, Ser^{204}, Ser^{207} in tm5 and Phe^{290} in tm6. The movement of Tyr^{316} after agonist binding towards Asp^{79} may be important for signal transmission to G_s. Whether all the interactions with the ligand occur simultaneously or sequentially is not known (modified from ref. 1).

in marked attenuation of agonist-stimulated rapid phosphorylation and receptor desensitization.

4.10. PERSPECTIVES

4.10.1 Additional Subtypes

The initial rapid expansion of the number of adrenergic receptor subtypes identified by molecular cloning has led scientists to wonder how many more would be found. A lucid analysis of the literature has now in fact matched most of the known pharmacological data with the properties of the cloned receptors expressed in model systems, suggesting that most adrenergic receptor subtypes have now been cloned. Remaining discrepancies have been linked to species differences rather than to evidence for more receptor subtypes: e.g. reports that the human β_3AR is pharmacologically different from the rat β_3AR receptor and may therefore represent yet another receptor subtype have not been borne out when sufficient numbers of ligands were tested in several homologous well controlled testing systems. Single residue differences between rat and human β receptors explained variations in pharmacologic properties, proposed by some investigators to reflect the existence of different subtypes. Finally, the ability of α-AR specific compounds to interact also with putative non-adrenergic imidazoline receptors may also explain apparent discrepancies in receptor properties.

4.10.2 Future Research

The molecular characterization of three β-adrenergic receptor subtypes now provides ample opportunities for future research. The comparison of homologous proteins which bind the same natural agonists with comparable affinities but synthetic agonists or antagonists with widely varying affinities will allow exquisitely precise structure-activity relationship studies. Site-directed mutagenesis, affinity labeling analysis and ultimately X-ray diffraction of the crystallized proteins should help establish a well-defined picture of the ligand binding site and elucidate changes in conformation leading to signal transmission and G protein activation. These analyses may also help in our understanding of the impact of some amino acid replacements on development of disease such as obesity for the β_3AR Trp $\rightarrow$ Arg mutation.

By completing these studies with the definition of the pharmacophores adapted to each subtype, selective drugs will become available to serve as therapeutic agents. This, however, will require the accurate definition of the physiologic function of each subtype which will first depend on the exact tissue localization by in situ hybridization or by antibody detection. The striking evidence for differential subtype regulation at the protein and mRNA levels also offers new avenues for future therapeutic approaches which act on gene expression rather than directly on receptor function itself.

ACKNOWLEDGMENTS

Support for our work comes mostly from the Centre National de la Recherche Scientifique, the Institut National de la Santé et de la Recherche Médicale, the University of Paris, the Ministry for Education and Research. We are also grateful for help from the Ligue Nationale contre le Cancer, the Fondation pour la Recherche Médicale Française and last but not least, the Association pour la Recherche contre le Cancer.

REFERENCES

1. Strosberg AD. Structure, function and regulation of adrenergic receptors. Protein Sci 1993; 1198-1209.
2. Ahlquist RP. A study of adrenotropic receptors. Am J Physiol 1948; 153:586-600.
3. Lands AM, Arnold A, MacAuliff JP, Luduena FP, Brown Jr TG. Differentiation of receptor systems activated by sympathomimetic amines. Nature 1967; 214:597-598.
4. Bylund DB, Blaxall, HS, Iversen LJ, Caron MG, Lefkowitz RJ, Lomasney JW. Pharmacological characteristics of β2-adrenergic receptors: comparison of pharmacologically defined subtypes with subtypes identified by molecular cloning. Mol Pharmacol 1992A; 42:1-5.
5. Bylund DB. Subtypes of β1- and β2-adrenergic receptors FASEB J 1992b; 6:832-839.
6. Emorine LJ, Marullo S, Briend-Sutren, MM, Patey G, Tate K, Delavier-Klutchko C, and Strosberg AD. Molecular characterization of the human β_3-adrenergic receptor. Science 1989; 245:1118-1121.
7. Arch JRS, Kaumann AJ. β_3 and atypical β-adrenoceptors. Med Res Rev 1993; 13:663-729.
8. Zaagsma J, Nahorski SR. Is the adipocyte β-adrenoceptor a prototype for the recently cloned atypical β_3-adrenoceptor? Tr Pharmacol Sci 1990; 11:3-7.
9. Vauquelin G, Geynet P, Hanoune J, Strosberg AD. Isolation of adenylate cyclase-free β-adrenergic receptor from turkey erythrocytes membranes by affinity chromatography. Proc Natl Acad Sci USA 1977; 74:3710-3714.
10. Vauquelin G, Geynet P, Hanoune J, Strosberg AD. Isolation of adenylate cyclase-free β-adrenergic receptor from turkey erythrocytes. Eur J Biochem 1979a; 98:543-556.
11. Caron MG, Srinivasan Y, Pitha J, Kiolek K, Lefkowitz RJ. Affinity chromatography of the β-adrenergic receptor. J Biol Chem 1979; 254:2923-2927.
12. Dixon RAF, Sigal IS., Candelore MR, Register RB, Scattergood W. Structural features required for ligand binding to the β-adrenergic receptor. EMBO J 1987; 6:3269-3275.
13. Yarden R, Rodriguez H, Wong SKF, Brandt DR, May DC, Burnier J, Harkins RN, Chen EY, Ramachandran J, Ullrich A, Ross EM. The avian β-adrenergic receptor: primary structure and membrane topology. Proc Natl Acad Sci USA 1986; 83:6795-6799.

14. Kobilka BK, Matsui H, Kobilka TS, Yang-Feng TL, Francke U, Caron MG, Lefkowitz RJ, Regan JW. Cloning, sequencing and expression of the gene coding for the human platelet β2-adrenergic receptor. Science 1987a; 238:650-656.
15. Henderson R, Baldwin JM, Cesca TAK, Zemlin F, Beckmann E, Downing KH. Model for the structure of bacteriorhodopsin based on high-resolution electron cryo-microscopy. J Mol Biol 1990; 213:899-929.
16. Dohlman HG, Thorner J, Caron MG, Lefkowitz RJ. Model systems for the study of seven-transmembrane segment receptors. Annu Rev Biochem 1991; 60:653-688
17. Wong SKF, Claughter C, Ruoho AE, Ross EM. The catecholamine binding site of the β-adrenergic receptors is formed by juxtaposed membrane-spanning domains. J Biol Chem 1988; 263:7925-7928.
18. Emorine LJ, Marullo S, Delavier-Klutchko C, Kaveri SV, Durieu-Trautmann O, Strosberg AD. Structure of the gene for the human β_2-adrenergic receptor: Expression and promoter characterization. Proc Natl Acad Sci USA 1987; 84:6995-6999.
19. Kobilka BK, Dixon RAF, Frielle T, Dohlman HG, Bolanowski MA, Sigal IS, Yang-Feng TL, Francke U, Caron, Lefkowitz RJ. cDNA for the human β_2-adrenergic receptor: a protein with multiple membrane-spanning domains are encoded by a gene whose chromosomal location is shared with that of the platelet-derived growth factor. Proc Natl Acad Sci USA 1987b; 84:46-50.
20. Frielle T, Collins S, Daniel KW, Lefkowitz RJ, Kobilka BK. Cloning of the gene for the β_1-adrenergic receptor. Proc Natl Acad Sci USA 1987; 84:7920-7924.
21. Emorine LJ, Marullo S, Briend-Sutren, MM, Patey G, Tate K, Delavier-Klutchko C, and Strosberg AD. Molecular characterization of the human β_3-adrenergic receptor. Science 1989; 245:1118-1121.
22. Grannenman JG, Lahners KN, Rao DD Rodent and human β3-adrenergic receptor genes contain an intron within the protein-coding block Molecular Pharmacology 1992; 42:964-970.
23. Van Spronsen A, Nahmias C, Krief S, Briend-Sutren MM, Strosberg AD, Emorine LJ. The human and mouse β_3-adrenergic receptor genes: promoter and intron/exon structure. Eur J Biochem 1993; 213:1117-1124.
24. George CC, Arbabian MA, Ruoho AE, Kiely J, Malbon CC. High efficiency expression of mammalian β-adrenergic receptor in baculovirus-infected insect cells. Biochem. Biophys. Res. Com. 1990; 163:1265-1269.
25. Ravet V, Blin N, Guillaume J-L, Petitjean F, Cabanie L, Strosberg AD. High level functional expression of human β_1-adrenergic receptor in baculovirus-infected cells screened by a rapid in situ procedure. J Receptor Res 1993; 13:541-558.
26. Mouillac B, Caron M, Bonin H, Dennis M, Bouvier M. Agonist-modulated pamitoylation of β_2-adrenergic receptor in Sf9 cells. J Biol Chem 1992; 267:21733-21737.
27. Marullo S, Delavier-Klutchko C, Guillet JG, Charbit A, Strosberg AD,

Emorine LJ. Expression of human β_1 and β_2-adrenergic receptors in *E. coli* as a new tool for ligand screening. Bio/Technology 1989; 7:923-927.
28. Marullo S, Delavier-Klutchko C, Eshdat Y, Strosberg AD, Emorine L. Human β_2-adrenergic receptors expressed in *Escherichia coli* membranes retain their pharmacological properties. Proc Natl Acad Sci USA 1988; 85:7551-7555.
29. Strosberg AD, Marullo S. Functional expression of G protein-coupled receptors in microorganisms. Tr Pharmacol Sci 1992; 13:95-98.
30. Breyer RM, Strosberg AD, Guillet JG. Mutational analysis of ligand binding activity of β_2-adrenergic receptor expressed in *Escherichia coli.* EMBO J 1990; 9:2679-2684
31. Marullo S, Emorine LJ, Strosberg AD, Delavier-Klutchko C. Selective binding of ligands to β_1/β_2 or chimeric $\beta 1/\beta 2$-adrenergic receptors involves multiple sub sites. EMBO J 1990; 9:1471-1476.
32. Bertin B, Freissmuth M, Breyer R, Schütz W, Marullo S, Strosberg AD. Functional Expression of the human serotonin 5HT-1A Receptor in *Escherichia coli:* Ligand binding properties and interaction with recombinant G protein α-subunits. J Biol Chem 1992; 267:8200-8206.
33. Grisshammer R, Duckworth R, Henderson R. Expression of a rat neurotensin receptor in *Escherichia coli.* Biochem J 1993; 295:571-576.
34. Grisshammer R, Little J, Aharony D. Expression of rat NK-2 neurokinin A; receptor in *E. coli.* Receptors and Channels 1994; 2:295-302.
35. Freissmuth M, Selzer E, Marullo S, Schutz W, Strosberg AD. Expression of two human β-adrenergic receptors in *E. coli:* functional interaction, with two forms of Gs. Proc Natl Acad Sci USA 1991; 88:8548-8552.
36. King K, Dohlman HG, Thorner J, Caron MG, Lefkowitz RJ. Control of yeast mating signal transduction by a mammalian β_2-adrenergic receptor and $G_s\alpha$ subunit. Science 1990; 250:121 123.
37. Green SA, Turki J, Innis M, Liggett SB Amino-terminal polymorphisms of the human β_2-adrenergic receptor impart distinct agonist promoted regulatory properties. Biochemistry 1994; 33:9414-9419.
38. Green SA, Cole G, Jacinto M, Innis M, Liggett SB. A polymorphism of the human β_2-adrenergic receptor within the fourth transmembrane domain alters ligand binding and functional properties of the receptor. J Biol Chem 1993; 268:23116-23121.
39. Walston J, Silver K, Bogardus C, Knowler WC, Celi FS, Austin S, Manning B, Strosberg AD, Stern MP, Raben N, Sorkin JD, Roth J, Shuldiner AR. Time of onset of non-insulin-dependent diabetes mellitus and genetic variation in the β_3-adrenergic receptor gene. New Engl J Med 1995; 333:343-347.
40. Clément K, Vaisse C, Manning B, Basdevant A, Guy-Grand B, Shuldiner A, Froguel P, Strosberg AD. Genetic variation in the β_3-adrenergic receptor and an increased capacity to gain weight in patients with morbid obesity. New Engl J Med 1995; 333:352-354.
41. Widen E, Lehto M, Kanninen T, Walston J, Shuldiner AR, Groop LC. Association of a polymorphism in the β_3-adrenergic receptor gene with

features of the insulin resistance syndrome in Finns. New Engl J Med 333:348-351.
42. Bertin B, Mansier P, Makeh I, Briand P, Rostene W, Swynghedauw B, Strosberg AD. Specific atrial over-expression of G protein-coupled human β_1-adrenergic receptors in transgenic mice. Cardiovascular Res 1993; 27:1606-1612.
43. Mansier P, Médigue C, Charlotte N, Vermeiren C, Coraboeuf E, Deroubai E, Clairambault J, Carré T, Dakhli T, Bertin B, Briand P, Strosberg AD, Swynguedauw B. Decrease in heart rate variability in transgenic mice over expressing atrial β_1-adrenoceptors. Science 1995; submitted.
44. Milano CA, Allen LF, Rockman HA, Dolber PC, McMinn TR, Chien KR, Johnson TD, Bond RA, Lefkowitz RJ. Enhanced myocardial function in transgenic mice overexpressing the β_2-adrenergic receptor. Science 1994; 264:582-586.
45. Rubenstein RC, Wong SKF, Ross EM. The hydrophobic tryptic core of the β-adrenergic receptor retains Gs regulatory activity in response to agonists and thiols. J Biol Chem 1987; 262:16655-16662.
46. Manning B St J, Pietri-Rouxel F, Drumare MF, Strosberg AD. Site-directed mutagenesis of human β_3-adrenergic receptor: Analysis of signal transduction mechanisms. manuscript in preparation.
47. Galzi-Méjean A, Guillaume JL, Strosberg AD. Carazolol: a potent, selective β3-adrenoceptor agonist. Eur J Pharmacol 1996; in press.
48. Tota MR, Strader CD. Characterization of the binding domain of the β-adrenergic receptor with the fluorescent antagonist carazolol. J Biol Chem 1990; 265:16891-16897.
49. Strader CD, Candelore MR, Hill WS, Dixon RAF, Sigal IS. A single amino acid substitution in the β-adrenergic receptor promotes partial agonist activity from antagonists. J Biol Chem 1989a; 264:16470-16477.
50. Strader CD, Candelore MR, Hill WS, Dixon RAF, Sigal IS. Identification of two serine residues involved in agonist activation of the -adrenergic recptor. J. Biol. Chem. 1989b; 264:13572-13578.
51. Strader CD, Dixon RAF, Cheung AH, Candelore MR, Blake AD, Sigal IS. Mutations that uncouple the β-adrenergic receptor from G_s and increase agonist affinity. J Biol Chem 1987a; 262:16439-16443.
52. Strader CD, Sigal IS, Blake AD, Cheung AH, Register B, Rands E, Zemcik B, Candelore MR, Blake AD, Dixon RAF. The carboxyl terminus of the hamster β-adrenergic receptor expressed in mouse L cells is not required for receptor sequestration. Cell 1987b; 49:855-863.
53. Fraser CM, Chung FZ, Wang CD, Venter JC. Site-directed mutagenesis of human β-adrenergic receptors: substitution of aspartic acid-130 by asparagine produces a receptor with high affinity agonist binding that is uncoupled from adenylyl cyclase. Proc Natl Acad Sci USA 1988; 85:5478-5482.
54. Cotecchia S, Ostrowski J, Kjelsberg MA, Caron MG, Lefkowitz RJ. Discrete amino acid sequences of the βl-adrenergic receptor determine the selectivity of coupling to phosphatidylinositol hydrolysis. J Biol Chem 1992; 267:1633-1639.

55. Suryanaryana S, Daunt DA, von Zastrow M, Kobilka BK. A point mutation in the seventh hydrophobic domain of the β2-adrenergic receptor increases its affinity for a family of β receptor antagonists. J Biol Chem 1992; 266:15488-15492.
56. Cotecchia S, Exum S, Caron MG, Lefkowitz RJ. Regions of the of β1-adrenergic receptor involved in coupling to phosphatidyl inositol hydrolysis and enhanced sensitivity of biological function. Proc Natl Acad Sci USA 1990; 87:2896-2900.
57. Samama P, Cotecchia S, Costa T, Lefkowitz RJ. A mutation-induced activated state of the β_2-adrenergic receptor. J Blot Chem 1993; 268:4625-4636.
58. Zilberfarb V, Piétri-Rouxel F, Krief S, Strosberg AD. Conversion of human immortalized brown pre-adipocytes into adipocytes leads to expression of functional β_3-adrenergic receptors. In preparation.
59. Collins S, Caron MG, Lefkowitz RJ. From ligand binding to gene expression: new insights into the regulation of G-protein coupled receptors. Tr Pharmacol Sci 1992; 17:37-39.
60. Thomas RF, Holt BD, Schwinn DA, Liggett SB. Long-term agonist exposure induces up-regulation of β_3AR exposure via multiple cAMP response elements. Proc Natl Acad Sci USA 1992; 89:4490-4494.
61. Hausdorff WP, Bouvier M, O'Dowd BF, Irons GP, Caron MG, Lefkowitz RJ. Phosphorylation sites on two domains of the human β_2-adrenergic receptor are involved in distinct pathways of receptors desensitization. J Biol Chem 1989; 265:12657-12665.
62. Lohse MJ, Lefkowitz RJ, Caron MG, Benovic JL. Inhibition of β-adrenergic receptor kinase prevents rapid homologous desensitization of β_2-adrenergic receptors Proc Natl Acad Sci USA 1989; 86:3011-3015.
63. Benovic JL, Major F. Jr, Staniszewski C, Lefkowitz RJ, Caron MG. Purification and characterization of the β-adrenergic receptor kinase. J Biol Chem 1987; 262:9026-9032.
64. Lohse MJ, Benovic JL, Codina J, Caron MG, Lefkowitz RJ. β-arrestin: a protein that regulates β-adrenergic receptor function. Science 1990; 248:1547-1550.
65. Strosberg AD. Biotechnology of the β-adrenergic receptors. Molecul Neurobiol 1991; 4:211-250.

CHAPTER 5

Metabotropic Glutamate Receptors: 10 Years of Research

Joël Bockaert and Jean-Philippe Pin

5.1 INTRODUCTION

In the brain, glutamate (Glu) is the main neurotransmitter used to trigger fast excitatory synaptic transmission. This is produced by activation of ligand-gated channels called ionotropic Glu receptors.[1-5] These are cation-specific ion channels and are sub-divided into N-methyl-D-aspartate (NMDA), α-amino-3-hydroxy-5-methyl-isoxazole-4-propionate (AMPA) and kainate (KA) receptors. They were the only receptors studied ten years ago when we described in 1985, that Glu and quisqualate (Quis) stimulated phospholipase C (PLC) in cultured striatal neurons.[6] The receptor involved was pharmacologically distinct from ionotropic receptors. Soon after, a similar effect of Glu was described in hippocampal slices,[7,8] cultured cerebellar granule cells[9] and cultured astrocytes[10] by other groups. These results strongly suggested that Glu, like GABA, serotonin and acetylcholine, not only activated ligand-gated channel receptors but also receptors coupled to GTP-binding proteins (G-proteins). The existence of such novel Glu receptors, now called metabotropic Glu receptors (mGluRs), was confirmed using the *Xenopus* oocyte model[11] and new pharmacological tools.[11,12]

In parallel with these studies, another non-ionotropic Glu receptor was characterized in brain and retinal bipolar cells, using the L-2-amino-4-phosphonobutyrate (L-AP_4) as specific ligand.[13-15] It was recognized to be coupled to G-proteins and therefore to be a mGluR, in 1990.[14]

cDNA coding for a mGluR coupled to PLC and having a pharmacology similar to the one we described in striatal cells[6] was cloned

Recombinant Cell Surface Receptors: Focal Point for Therapeutic Intervention,
edited by Michael J. Browne.

4 years ago,[15-16] whereas a first L-AP_4 sensitive mGluR was cloned in 1992.[17] The cloning effort reveals that the mGluR family consists of at least eight subtypes: $mGluR_1$ to $mGluR_8$ (Fig. 5.1).[17-21] Some mGluRs RNA are subjected to alternative splicing so that different mGluR variants are generated by a single gene. To date, 3 splice variants have been identified for $mGluR_1$ ($mGluR_1a$, $mGluR_1b$ and $mGluR_1c$),[17,22] 2 for $mGluR_5$ ($mGluR_5a$ and $mGluR_5b$),[23,24] and 2 for $mGluR_4$ for 2 ($mGluR_4a$ and $GluR_4b$).[25,26]

Recent important work on mGluRs includes the identification of more specific agonists and antagonists of the different subtypes,[27,28] the production of mutant mice lacking $mGluR_1$[29-31] and $mGluR_6$,[32] as well as the demonstration of physiological roles for mGluRs.[27,33] This chapter provides a brief outline of our knowledge on the structure, the pharmacology, the transduction mechanisms, and the physiological and pathological roles of mGluRs. For other recent reviews see refs. 27, 28, 33, 34.

5.2 THE mGluR FAMILY AND THE UNIQUE STRUCTURE OF mGluR PROTEIN COMPARED TO OTHER G-PROTEIN COUPLED RECEPTORS (GPCRs)

Based on their amino acid sequence identity, the 8 mGluRs can be classified into 3 groups (Fig. 5.1). mGluRs of the same group show about 70% sequence identity, whereas between groups this percentage falls to about 45% (Fig. 5.1). Group I comprises $mGluR_1$ and $mGluR_5$, group II, $mGluR_2$ and $mGluR_3$, group III, mGluRs 4,6,7 and 8. Recently, a Ca^{2+}-sensing receptor, isolated from a bovine parathyroid cDNA library, has been found to have about 30% sequence identity with mGluRs.[35] This receptor clearly belongs to the mGluR family (Fig. 5.1).

All the GPCRs possess seven putative transmembrane domains. However, despite the overall stucture similarity with other GPCRs, the mGluR family (family 3; Fig. 6.2) seems to be distinct. It shows no sequence similarity with the two other GPCR families, i.e. the rhodopsin receptor family (family 1, which includes most of the GPCRs (Fig. 5.2) on one hand, and the secretin/glucagon/pituitary adenylyl cyclase activating polypeptide (PACAP)/vasoactive intestinal peptide (VIP) receptor family (family 2; Fig. 5.2)[36] on the other hand. No sequence homology exists between the three families. The superfamily of GPCRs may thus be subdivided into three subgroups (as proposed in Figure 6.2) which probably derives from three distinct ancestral genes. The presence of seven transmembrane domains may result from a convergent evolution for these proteins, to interact optimally with common G-proteins.

As mentioned above, splice variants have been found for three mGluRs: $mGluR_1$, $mGluR_4$ and $mGluR_5$.[17,22-24,26] The splice variants result from the use of alternative acceptor splice sites in an intron whose position is conserved at least between these three genes. In

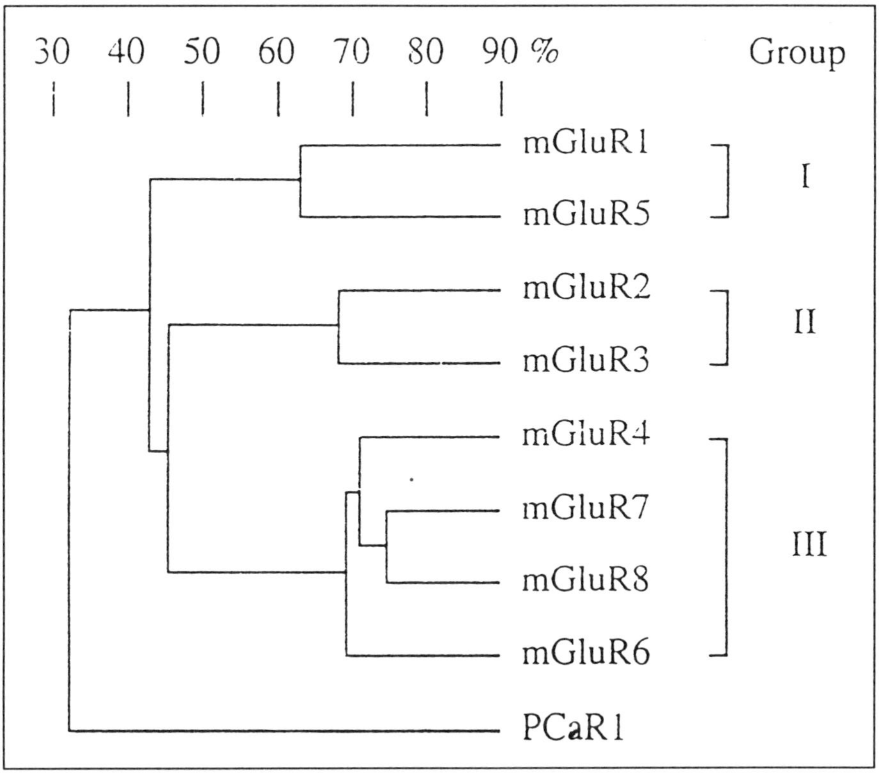

Fig. 5.1. Dendrogram and pharmacological classification of mGluR members including the bovine parathyroid Ca^{2+}-sensing receptor (PCaR). The numbers on the top indicate the % amino acid sequence identity between members of this receptor family (at the junction of the horizontal lines).

mGluR$_1$b, the insertion of an additional 85 base pair exon, which contains an in frame stop codon, results in the deletion of 318 amino acids from the proposed cytoplasmic carboxy-terminal domain and addition of 20 different residues at the carboxy-terminal tail.[17] In mGluR$_1$c, a distinct insertion results in a similar deletion and addition of 11 other amino acids at the carboxy-terminus.[22] The mGluR$_5$b variant results from an in frame insertion of 96 bases, and thus 32 amino acids are inserted into the deduced protein.[23,24] For mGluR$_4$, the first cloned form mGluR$_4$a, possesses an additional exon of 621 bases containing an in frame stop codon; deletion of this exon in mGluR$_4$b results in a longer protein in which the last 63 C-terminal residues are replaced by 136 residues.[25,26]

5.2.1 The Ligand Binding Site

The agonist binding site of the GPCR family 1 has been extensively studied. In the case of small ligands, such as catecholamines, it is located in a pocket formed by the seven transmembrane domains (family 1a; Fig. 5.2).[37-39] In the case of some peptides and cytokines, the binding site includes both the N-terminal domain and the external parts of transmembrane domains (family 1b; Fig. 5.2).[40] In glycoprotein receptors (LH, FSH...) (family 1c; Fig. 5.2) as well as in family 2 receptors (Fig. 5.2) the binding site is essentially (perhaps not entirely) in the N-terminal extracellular domain.[41,42]

Two studies identify the mGluR agonist binding site in the large amino-terminal extracellular domain of about 600 residues (Fig. 5.3).[43,44]

Takahashi et al[44] took advantage of the different pharmacological profiles of $mGluR_1$ and $mGluR_2$ to map the regions involved in the agonist selectivity. By exchanging parts of the extracellular domains of these two receptors, they showed that the two thirds of this domain are sufficient to convert the pharmacological profile of $mGluR_1$ into that of $mGluR_2$. Computational analysis by O'Hara et al[43] revealed that this domain shares sequence homology with bacterial periplasmic proteins (BPP) that bind amino acids or ions. The crystal structure of these bacterial proteins was therefore used to construct a model for the glutamate binding site in $mGluR_1$.[43] This model predicted that the 500 N-terminal residues form 2 globular lobes with a hinge region where glutamate binds (Fig. 5.3), and these two lobes are proposed to close like a clamshell upon agonist binding. This model is supported by mutagenesis of serine 165 and threonine 188 of $mGluR_1$ affecting the glutamate affinity.[43] It is noteworthy that mutations in the close proximity of the hinge region of the parathyroid Ca^{2+} sensing receptor have been identified in patients suffering from genetic diseases affecting Ca^{2+} homeostasis.[45,46] Finally, this model recently received indirect confirmation from work on the glutamate receptor channel subunits.[28,47-49]

5.2.2 G-Protein Coupling

In contrast to family 1 and 2 of GPCRs,[39,50] it is not the third intracellular loop (i3; Fig. 5.3) which governs the specificity of the

Fig. 5.2. (opposite page) Classification of G protein coupled receptors (GPCRs) Family 1: this family contains the majority of GPCRs. One of their characteristics is the DRY(aspartyl-arginyl-tyrosyl) or EWR (glutamyl-arginyl-tryptophanyl) sequences in the N-terminal region of i2 as well as highly conserved amino acids in transmembrane domains (such as an aspartyl(D) in the transmembrane II). This group can be divided into 3 sub-families; subfamily 1a *which comprises most receptors for small ligands and in which the binding site is localized within the transmembrane domains.* subfamily 1b *which comprises many receptors for peptides. The binding site is localized between the extracellular N terminal domain and the external parts of the transmembrane domains.* subfamily 1c *which comprises receptors for glycoprotein hormones. The binding site is mainly localized in the extracellular N terminal domain. Family 2: which comprises receptors for secretin, glucagon, PACAP etc. in which the binding site lies essentially in the extracellular N-terminal domain. Family 3: which comprises mGluRs and Ca^{2+} sensing receptors. There is no homolgy between the three families which probably derived from separate genes.*

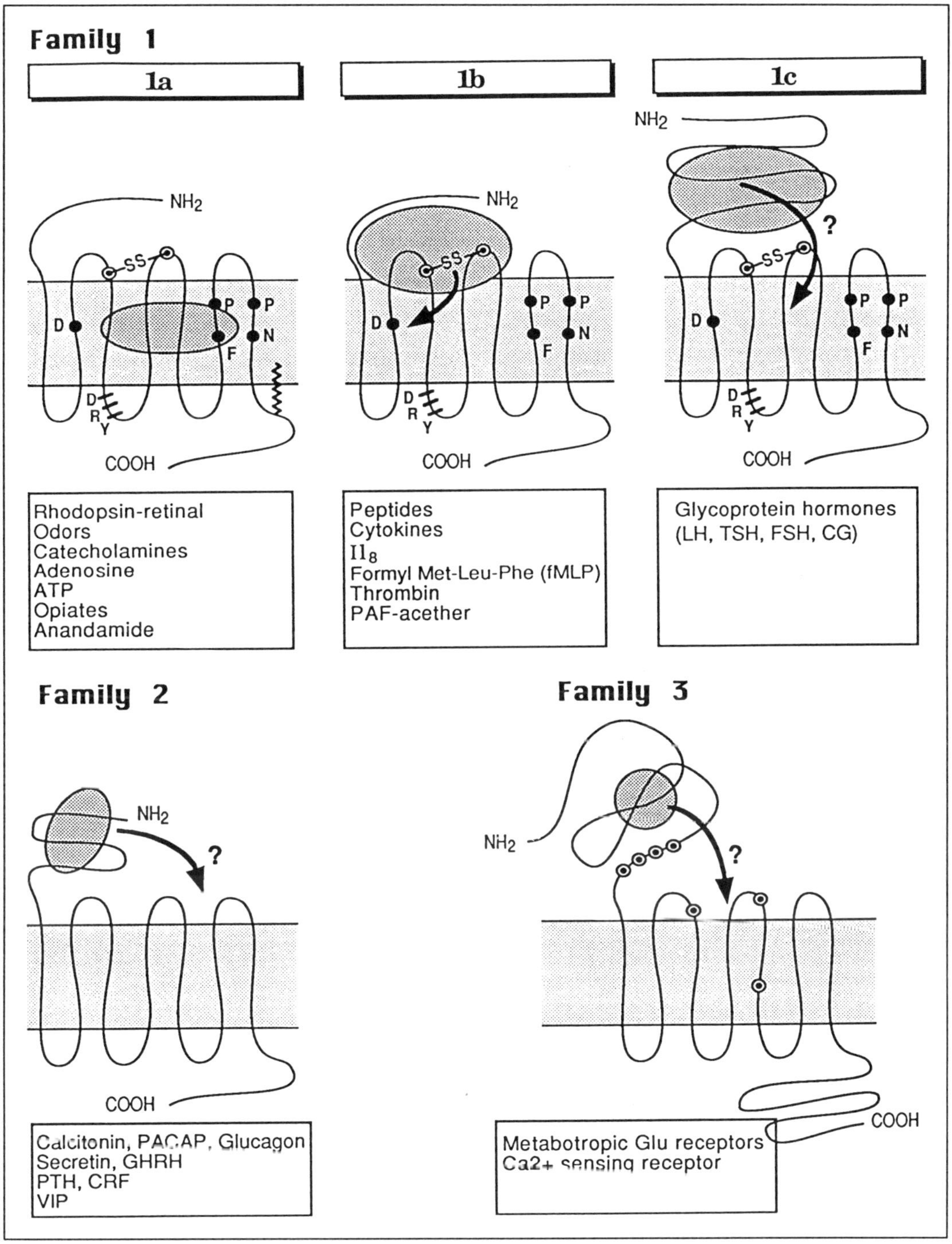
Family 1
1a
1b
1c
NH2
SS
D
P
N
F
R
Y
COOH
?
Rhodopsin-retinal
Odors
Catecholamines
Adenosine
ATP
Opiates
Anandamide
Peptides
Cytokines
Il8
Formyl Met-Leu-Phe (fMLP)
Thrombin
PAF-acether
Glycoprotein hormones
(LH, TSH, FSH, CG)
Family 2
Calcitonin, PACAP, Glucagon
Secretin, GHRH
PTH, CRF
VIP
Family 3
Metabotropic Glu receptors
Ca2+ sensing receptor

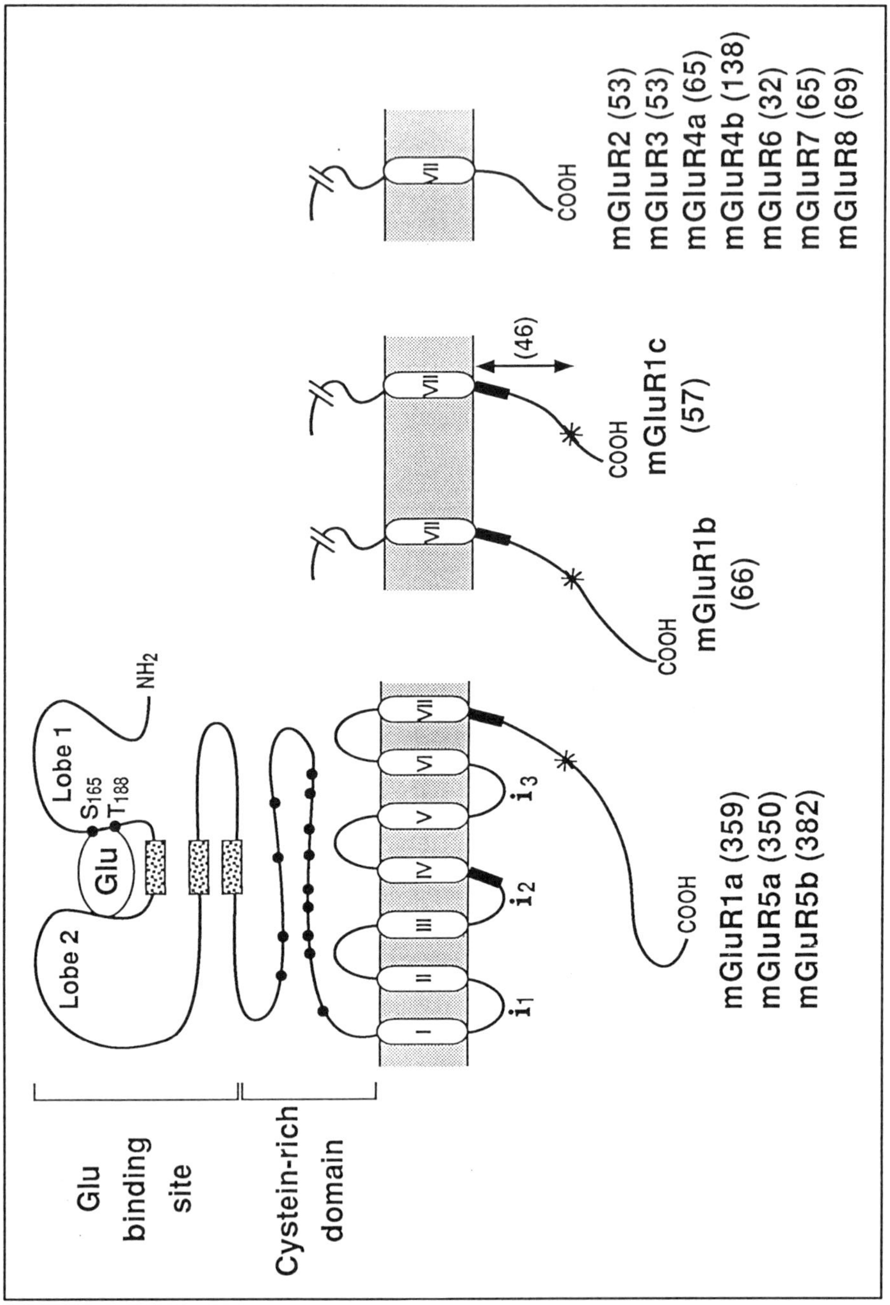

Fig. 5.3. Schematic representation of mGluR$_1$a. The putative Glu binding domain is localized between two lobes. In this region, two amino acids involved in Glu binding are indicated. The cysteine residues conserved among all mGluRs are indicated by black circles. The position of the intron of the mGluR$_1$ gene where alternative splicing occurs is indicated by an asterisk. The domains involved in the specificity of G-protein activation are represented with thick black lines. The number in parenthesis is the number of amino-acids in the C-terminal domain (C-t) of the different mGluRs.

coupling to the G-proteins, but the second intracellular loop (i2; Fig. 5.3) and more precisely the 16 C-terminal residues of i2.[51] Alone, however, this domain is not sufficient to obtain coupling to G protein and additional intracellular domains are needed, either a minimal stretch of 28 to 57 residues of the amino-terminal portion of the carboxy-terminal (C-t) domain,[51] or another intracellular loop (i1 or i3).[52] Note that the C-t is more efficient than any other intracellular loop in this regard. The length of the C-t also regulates qualitatively and quantitatively coupling to G-proteins. Indeed, $mGluR_1a$ (C-t=359 residues; Fig. 5.3) and $mGluR_5$ (C-t=350 residues; Fig. 5.3) display intrinsic activity when transfected in cell lines and increase cAMP production, whereas $mGluR_1b$ (C-t = 66 residues; Fig. 5.3) and $mGluR_1c$ (57 residues; Fig. 5.3) are devoid of such activity[23] (and Prézeau et al in preparation). In addition, $mGluR_1a$, $mGluR_5a$ and $mGluR_5b$ give a sharp transient response in *Xenopus* oocytes, whereas $mGluR_1b$ and $mGluR_1c$ give slowly generated responses.[22,23,52] Finally, although part of i2 plus a short part of the C-t is enough to trigger a potent coupling to G-proteins, experiments with chimeric receptors reveal that cooperation between all intracellular domains is necessary to obtain the full characteristics of native coupling to G-proteins of the different $mGluR_1$ spliced variants (Fig. 5.4).[52]

The structure of mGluRs raises the question of how a change in the structure of the protein in the extracellular domain can modify the structure of intracellular domains involved in G-protein coupling? One possibility is a secondary interaction of the binding domain (once Glu is locked) with the transmembrane domains, as already proposed for the family 1c of GPCRs (Fig. 5.2). The other possibility is to imagine a putative transducing role of the cysteine rich domain located between the binding site and the transmembrane domains (Fig. 5.3).

5.3 PHARMACOLOGY OF mGLURs

The focus of this section is on the pharmacological tools which can been used to recognize cloned mGluRs (Table 5.1) and have recently been reviewed.[27] Six available agonists can be used to characterize the different mGluR groups: 1) (1S,3R) trans-1-aminocyclopentane-1,3-dicarboxylate (ACPD); 2) Quis (in the presence of 6-cyano-7-nitroquinoxaline-2,3-dione (CNQX) to block ionotropic effects); 3) (2S,1'S,2'S)-2-(carboxycyclopropyl)glycine (L-CCG); 4) L-AP_4; 5) 2-(2,3-dicarboxycyclopropyl)glycine (DCG-IV); 6) (R,S) 3,5-dihydroxyphenylglycine (DHPG).

Groups I and II but not group III mGluRs are activated by ACPD. In contrast, group III but not groups I or II mGluRs are activated by L-AP_4.

In addition, group I is the only group fully activated by Quis at concentrations below 10 μM and by DHPG[53-57] and is unaffected by

Table 5.1. Pharmacological recognition of cloned mGluRs

	Discriminative agonists	Other Criteria
GROUP I		
	ACPD (FE) DHPG (FE) QUIS° (FE ≤ 10 μM) L - CCGI (NE) DCG - IV (NE) L - AP4 (NE)	mGluR1: antagonists (MCPG; 4 CPG; 4C 3HPG) mGluR5: MCPG (weak antagonist) ; 4 CPG (NE); 4C 3HPG (weak agonist)
GROUP II		
	ACPD (FE) L - CCGI (FE at 5 μM) DCG - IV* (FE at 5 μM) QUIS° (NE at 1 μM) L - AP4 (NE) DHPG (NE)	other agonists: 4C 3HPG; 4 CPG (at mGluR2) antagonists: MCPG (only tested on mGluR2); MCCG
GROUP III		
	L - AP4 (FE at 10 μM) (mGluR4, R6, R8) L - AP4 (FE at 2 mM) (mGluR7) ACPD (NE) L - CCGI (NE) DCG - IV (NE)	other agonist: L - SOP antagonist: MAP4

(°) tested in the presence of ionotropic Glu receptor antagonists
(*) stimulates NMDA receptors at high doses
FE: fully effective
NE: non effective

μM concentrations of L-CCG-I and DCG-IV[2,15,16,58,59] (Table 5.1). There is a debate as to whether 2-amino-3-phosphonopropionic acid (AP3), which has been shown to be an antagonist of group I mGluRs in many biological preparations (for discussion see ref. 27), is also an antagonist of cloned mGluRs coupled to PLC. Several reports indicate that it is not an antagonist.[18,60] A recent study using expression in *Xenopus* oocytes, however, indicates that it could in fact be a competitive inhibitor of group I cloned receptors.[61] Recently, it has been shown that $mGluR_1$ and $mGluR_5$ can be pharmacologically distinguished[53,62] using (S)-4-carboxy-phenylglycine (4CPG), which is a competitive antagonist at $mGluR_1$ and is almost inactive at $mGluR_5$.[53] To further distinguish between $mGluR_1$ and $mGluR_5$ one can also take into account the finding that (RS)-α-methyl-4-carboxyphenylglycine (MCPG) is more potent at $mGluR_1$ than at $mGluR_5$[53] and that (S)-4-carboxy-3-hydroxyphenylglycine (4C3HPG) which is a competitive antagonist at $mGluR_1$ is a weak agonist at $mGluR_5$.[53,62] Using these criteria, Manzoni et al have shown that presynaptic inhibition at the Schaffer/CA1 synapses is mediated by $mGluR_5$ and not by $mGluR_1$ or group II mGluRs.[57]

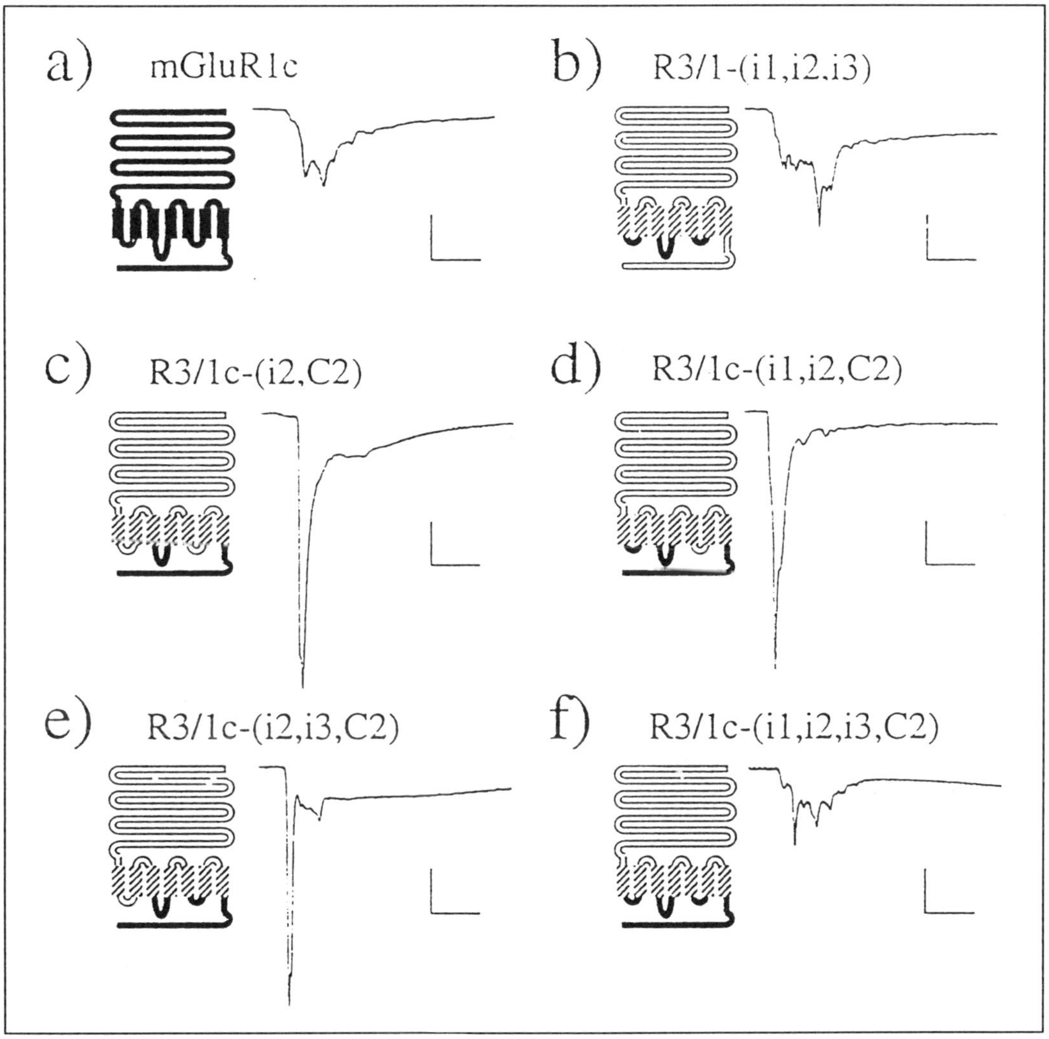

Fig. 5.4. Analysis of the kinetics of responses induced by $mGluR_1c$ and by the chimeric receptors made between $mGluR_1c$ and $mGluR_3$. The $mGluR_1c$ is designed in black and $mGluR_3$ in white. The intracellular loops i1, i2, i3 and the C-t of $mGluR_1c$ are introduced in $mGluR_3$. Only the introduction of the 3 intracellular loops plus (f) or minus (b) C-t gives a slow oscillating kinetic comparable with that observed with $mGluR_1c$ (a). The other chimeras show a kinetic response comparable with that observed with $mGluR_1a$ (see ref. 51).

In addition to ACPD, group II mGluRs are fully activated by μM concentrations of L-CCG-I and DCG-IV but unaffected by DHPG (but see ref. 63), L-AP_4 and 1 μM of Quis. The sensitivity of $mGluR_2$ and $mGluR_3$ to Quis is quite different, with an EC_{50} of 1 mM for $mGluR_2$ and 40 μM for $mGluR_3$, a difference which is enough to be useful in determining the exact mGluR subtype involved in a physiological response. In addition, 4CPG, 4C3HPG and 3H4CPG are agonists at $mGluR_2$.[2,51,58,59,64,65] The only known antagonist at group II mGluRs is MCPG (although so far it has only been tested on $mGluR_2$).[64] The α-methyl derivative of L-CCG-I (MCCG) found to be antagonist at group II mGluRs in physiological preparations,[66-68] has not yet been tested on cloned receptors.

As already mentioned, group III is the only group activated by L-AP_4 and which is insensitive to low concentrations of ACPD. However, L-AP_4 and Glu are far more potent at $mGluR_4$ and $mGluR_6$ than at $mGluR_7$.[20,21,65] L-serine-O-phosphate (L-SOP) is also an agonist of group III mGluRs. MCPG and the α-methyl derivative of L-AP_4 (MAP_4) has been found to be an antagonist at group III mGluRs in physiological preparations,[66-68] but their action on cloned group III receptors remain to be studied.

5.4 TRANSDUCTION MECHANISMS

This section will essentially summarize our knowledge on transduction mechanisms of cloned receptors (indicated by (c) in Table 5.2) and compares them with the transduction mechanisms determined for native mGluRs. For a more detailed review see ref. 27.

5.4.1 Group I

In every expression system examined, group I receptors, including their splice variants, stimulate PLC as revealed by an increase in phosphoinositide turnover, and Ca^{2+} release from internal stores. As already mentioned, some differences in the kinetics of Ca^{2+} release induced by $mGluR_1$a, b and c have been noticed in *Xenopus* oocytes and in transfected mammalian cells[22,25,52,69] (Fig. 5.4). This may reflect a difference in the affinity of the receptors for the G-protein. Among the group I receptors, $mGluR_1$a (but not $mGluR_1$b or $mGluR_1$c) as well as $mGluR_5$a and $mGluR_5$b have been shown to activate adenylyl cyclase when expressed in cell lines[23,60,70] (and Mary et al in preparation). Using on one hand our knowledge that $mGluR_1$ and $mGluR_5$ receptor mRNA are found in cerebellar granule cells and cortical neurons respectively,[71] and on the other hand, our demonstration that 4CPG is only active on $mGluR_1$,[23] we demonstrated that indeed, $mGluR_1$ and $mGluR_5$ stimulate PLC in cultured cerebellar granule cells and cortical neurons, respectively[53] (Table 5.2). $mGluR_1$a also activates phospholipase A_2 (PLA_2) when transfected in CHO cells, but the mechanism is unknown.[60] Similarly, a mGluR of group I has been shown to activate PLA_2 in striatal neurons when it is co-stimulated with AMPA

Table 5.2. Signal transduction of mGluRs

	PLC	PLD	PLA2	AC -	 +	VSCC activation N	 L	VSCC inhibition N	 L	IKM IKAHP	I leak	BK Channels	Non-Selective Cationic Channels	cGMP - PDE
GROUP I														
mGluR1 (c)	+		+											
mGluR5 (c)	+													
mGluR1 or R5	+		+				+	+	+	+		+		
mGluR1	+													
mGluR5	+													
GROUP II														
mGluR2 (c)				+		+								
mGluR3 (c)				+										
mGluR2 or R3				+	+(a)			+	+					
GROUP III														
mGluR4 (c)				+										
mGluR6 (c)				+										
mGluR7 (c)				+										
mGluR8 (c)				+										
mGluR6														+
mGluR4 like				+										
Non-characterized	+	+	+	+							+		+	

c: cloned receptors
a: in coincidence with C_s activating receptors

receptors.[72] The reversal of the Na^+/Ca^{2+} exchanger has been proposed to be the key step in this cooperative effect.[73]

Cytoplasmic injection of $mGluR_1a$ cRNA into sympathetic neurons results in expression of receptors that inhibit M-type K^+ channels.[74] Using pharmacological tools, it has been established that group I mGluRs can: 1) activate L-type voltage-sensitive Ca^{2+} channels (VSCC) in cerebellar granule cells[75,76] via an unknown messenger; and 2) inhibit either L or N-type VSCC in olfactory neurons,[77] cortical and pyramidal hippocampal neurons.[78-81] Although they involve a G-protein (not identified) the mechanisms of these inhibitions of VSCC by group I receptors are unclear. $mGluR_5$ probably inhibits synaptic transmission in the Schaffer/CA_1 synapses via inhibition of VSCC.[56,57]

Several subtypes of K^+-channels are regulated by mGluR agonists of group I (Table 5.2).[27] Inhibition of the voltage-dependent K^+-current IKM and the Ca^{2+}-activated K^+-current IK_{AHP} have been observed in hippocampal neurons[82-84] and neurons of the nucleus of the tractus solitarius.[85] In both cases, Quis is the most potent agonist, followed by ibotenate (Ibo) and ACPD. However, the phenylglycine derivatives 4C3HPG, 4CPG and MCPG have no effect on the ACPD-induced K^+-channel inhibition in neurons of the nucleus of the tractus solitarius.[85] Therefore it is possible that there could be another, as yet uncloned, mGluR (Table 5.2). The exact mechanism involved in this control of K^+-channel activity by mGluRs has not been clearly established. It is clear, however, that Ca^{2+}-activated K^+-current IK_{AHP} can be inhibited with no changes in Ca^{2+}_i.[83] Although PKC activation is apparently involved in the inhibition of IKAHP in dentate granule neurons, neither PKC nor PKA are involved in this effect in CA_3 pyramidal neurons.[84]

The activation of K^+-channels by group I mGluRs has also been observed. In cerebellar granule cells, a PLC-coupled-like mGluR activates the Ca^{2+}-activated K^+-channel called BK as a consequence of the increase in Ca^{2+}_i.[86]

5.4.2 Group II

When transfected into cell lines, $mGluR_2$ and $mGluR_3$ inhibit forskoline-stimulated cAMP production via a Pertussis toxin (PTX) sensitive G protein indicating that a G-protein of the G_i family is involved as is the case of other receptors inhibiting adenylyl cyclase.[17,65] Several groups have reported inhibition of adenylyl cyclase by mGluR agonists in brain slices, cultured neurons and cultured astrocytes (Table 5.2) (for a review see ref. 27). This Glu response is antagonized by PTX. The demonstration that Glu receptors directly coupled to adenylyl cyclase via inhibitory G-proteins exist in neurons was revealed by the observation that Glu inhibits adenylyl cyclase in plasma membranes.[71,87] Quis is a relatively good agonist in cortical and striatal neurons, and in cortical astrocytes, suggesting that $mGluR_3$ is involved. In contrast, Quis is almost inactive in cerebellar granule cells in agreement with its very low affinity on $mGluR_2$.[71,87] Injection of $mGluR_2$ cRNA into sympathetic neurons results in expression of receptors that inhibit N-type Ca^{2+} channels via a pertussis toxin G protein.[74] Group II mGluRs also inhibits the L and N type VSCC via a PTX sensitive G protein in cerebellar granule cells, likely via a soluble unknown messsenger.[75,88] Finally, in rat brain slices, several Glu analogues were found to potentiate cAMP formation which is stimulated by G_s coupled receptors.[89-92] A careful analysis of the pharmacology of the receptor involved indicates that it belongs to group II, although the potency of LCCG-I is relatively weak.[93] The mechanism involved is not clear, but may involve potentiation of αs stimulated adenylyl cyclase II by $\beta\gamma$ released by group II mGluRs from G_i or G_o proteins.[94]

5.4.3 Group III

Although group III mGluRs, inhibit cAMP production via a PTX sensitive G protein when transfected in cell lines, they are less potent than group II.[19,21,65,95,96] In guinea pig cortical slices,[97,98] rat cortical slices,[99] mice cortical and cerebellar granular cells in culture,[71] L-AP_4 also inhibits cAMP production. However, the receptors involved are not well characterized and in cortical slices it is inhibited by MCPG which does not antagonize $mGluR_4$ and $mGluR_7$, the two receptors expressed in this brain area. Therefore a new mGluR is perhaps involved (Table 5.2). In addition, in cerebellar granule cells, the maximal effect of L-AP_4 is very small and unrelated to the amount of $mGluR_4$ mRNA found in these cells.[65,71] Moreover, it was impossible to detect any inhibition of adenylyl cyclase with L-AP_4 on cerebellar granule cell membranes.[71] Thus, the exact transduction mechanism of $mGluR_4$ in neurons remains to be established. $mGluR_4$, like $mGluR_6$, may possibly be coupled to other G-proteins that regulate different effectors like cGMP phosphodiesterases. Group III mGluRs also inhibit VSCC of the L type in olfactory neurons.[77] It is likely that these receptors mediate inhibition of glutamic synaptic transmission in hippocampus.[57,67]

The L-AP_4 sensitive mGluR located on On-bipolar cells of the retina is $mGluR_6$. $mGluR_6$ is only expressed in those cells[19,100] and the response of On-bipolar cells disappear in $mGluR_6$ knock-out mice.[32] It has been proposed that $mGluR_6$ activates a cGMP-phosphodiesterase.[14,101] The proposed transduction cascade would therefore be identical to that following the activation of rhodopsin in photoreceptor cells: the activated G-protein stimulates a cGMP phosphodiesterase, leading to a decrease in cGMP concentration and the closure of cGMP-gated channels (Fig. 5.5) (see also section 5.5 for physiology).

5.4.4 mGluR Effects Mediated by Poorly Characterized Receptors

mGluRs having a pharmacology different from that of any of the cloned receptors have also been found to stimulate PLC in hippocampal

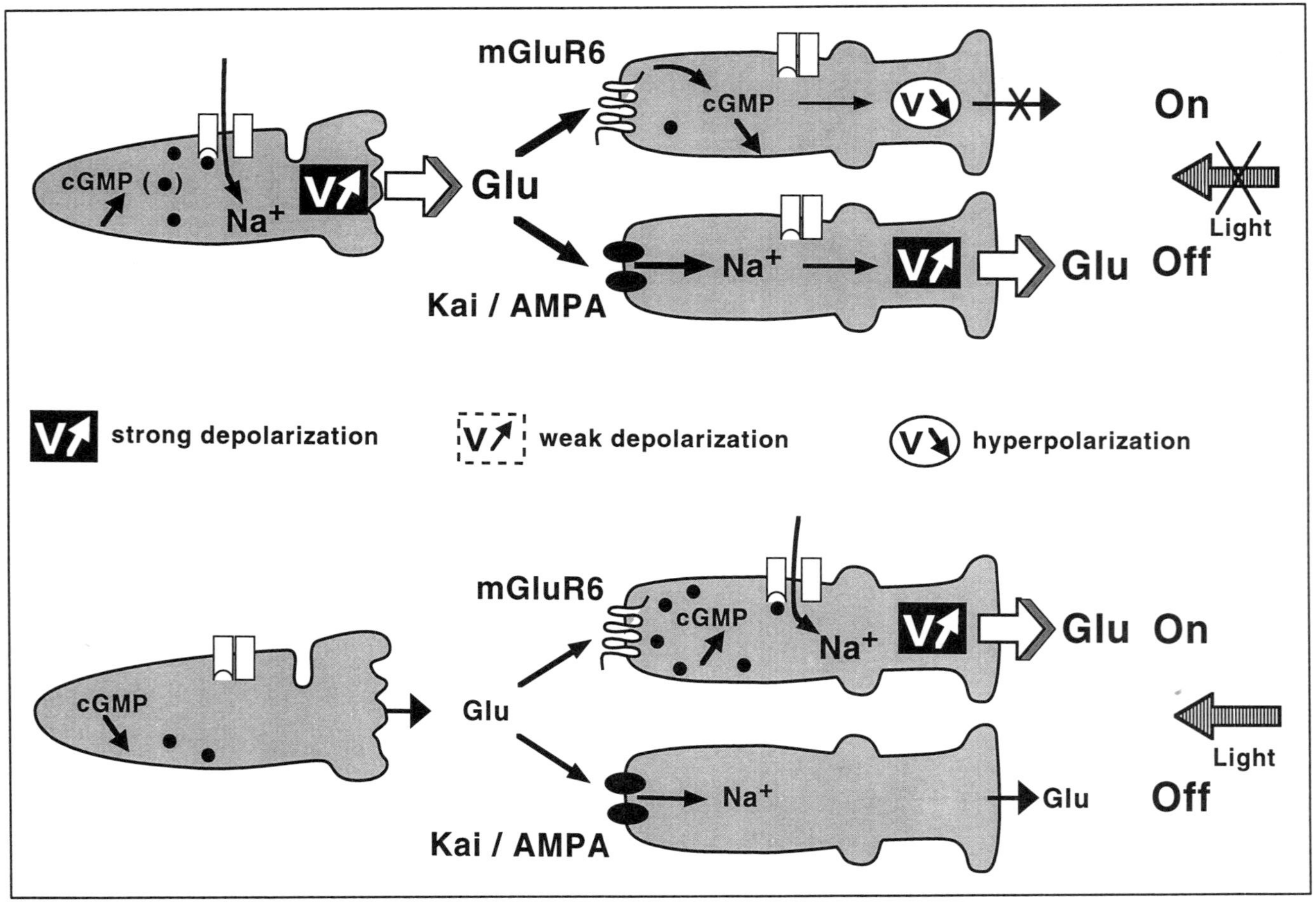

Fig. 5.5. Transduction mechanisms in photoreceptors and retinal bipolar cells. This model explains the synaptic mechanism for On and Off segregation and the function of $mGluR_6$.

slices,[102,103] PLD in hippocampal slices,[102-105] PLA_2 in astrocytes in culture,[106] or to inhibit cAMP accumulation in the same astrocytes.[71] Concerning the receptor which stimulates PLD, it is perhaps not a mGluR because it is stimulated by ACPD, L-cysteine sulfinic acid but not by glutamate.[105] We have already described the non identified group III, L-AP_4 sensitive mGluR which inhibits cAMP production in cortical slices.[99] Resting K^+-currents, called IKleak, can also be inhibited by 1S,3R-ACPD in CA3 hippocampal,[107] tractus solitarius nucleus[108] and dorsal lateral geniculate nucleus[109] neurons. Again, the pharmacological characterization of these effects is relatively poor. The mechanisms involved are also unknown.

Finally, a non identified mGluR, sensitive to ACPD and blocked by MCPG, has been found to activate a non selective cationic channel in rat hippocampus without implication of a G protein.[110]

5.5 PHYSIOLOGICAL ROLES OF mGluRs IN WILD AND MUTANT RODENTS

5.5.1 Synaptic Transmission

Synaptic transmission between photoreceptors and On bipolar cells is a specific case in which a mGluR ($mGluR_6$) has been shown to mediate glutamate transmission directly[33] (Fig. 5.5). When light activates photoreceptors, rhodopsin stimulates a cGMP phosphodiesterase via transducin, a G protein.[111] The decrease in cGMP hyperpolarizes photoreceptors by closing cGMP-gated channels and reduces Glu release. Under these conditions, $mGluR_6$ is not activated, the intracellular concentrations of cGMP are high in On-bipolar cells, the cGMP-gated channels of these cells are open maintaining depolarization and the release of Glu which excites the subsequent cell (Fig. 5.5). During the same period, the Off-bipolar cells are hyperpolarized because their AMPA/kainate receptors are not stimulated. During darkness, the symmetrical scheme operates. Thus, the role of $mGluR_6$ is to produce sign inversion in the visual transduction. As expected, the On-bipolar response to light is suppressed in $mGluR_6$ knock-out mice. However, the transmission to visual centers as well as the visual behavioral responses are unchanged[32] (Table 5.3).

A direct role for mGluRs in the slowly developing depolarization of Purkinje neurons after low frequency stimulation of parallel fibers,[112] as well as in the excitation of the ventro-basal thalamic neurons induced by noxious stimuli has been demonstrated.[113]

5.5.2 Modulation of Synaptic Transmission

We have already discussed the fact that mGluRs of the three groups are able to inhibit synaptic transmission of Glu synapses in the hippocampus.[57,67] This is not restricted to Glu synapses but mGluRs also inhibit inhibitory synapses such as GABA synapses (for review see ref. 27).

Table 5.3. mGluR knock - out mice

	mGluR1 Conquet et al 1994 (Ref. 31)	mGluR1 Aiba et al 1994 (Ref. 29-30)	mGluR6 Masu et al 1995 (Ref. 32)
Synaptic Transmission			
Schaffer collaterals/CA1	normal	normal	
Mossy fibers/CA3	normal		
Parallel fibers/Purkinje cells	normal	normal	
On bipolar cells			suppressed
Synaptic Plasticity			
LTP CA1	normal	reduced	
LTP CA3	impaired		
LTD CA1		normal	
LTD cerebellum	impaired	impaired	
Behavior			
Ataxia	yes	yes	
Spatial memory	impaired		
Associative memory		impaired	
Behavior of visual responses			normal

In the accessory olfactory bulb, immunohistochemistry revealed that $mGluR_2$ receptors are highly expressed in the granule cells where they are concentrated at the reciprocal synapses with mitral cells. At these synapses, $mGluR_2$ inhibits the release of GABA and therefore favors the excitatory transmission from the mitral to the granule cells.[59] Infusion of the $mGluR_2$ agonist DCG-IV in the accessory olfactory bulb of female mice allows them to memorize a male pheromone without mating, suggesting a role for this receptor in a form of olfactory memory.[114]

Note that in all these studies, inhibition of the synaptic transmission is obtained using a mGluR agonist, and the implication of mGluRs in physiological synaptic transmission remains to be conclusively established.

5.5.3 Synaptic Plasticity and Learning

The involvement of mGluRs at specific synapses has already revealed a role for these receptors in nociception,[113,115] olfactory memory (just discussed),[114] motor and spatial memory. In the cerebellum, LTD of parallel fiber Purkinje cells synapses is a well-defined phenomenon. A role for PLC-coupled mGluRs, in particular $mGluR_1$ that is highly expressed in Purkinje cells, has been proposed. This form of synaptic plasticity is thought to result from the co-activation of AMPA and mGluRs associated with an increase in Ca^{2+} induced by Purkinje cell depolarization. The use of $mGluR_1$ inactivating antibodies confirmed these models.[116] More recently, the generation of mice lacking the $mGluR_1$ gene was carried out independently by 2 groups (Table 5.3).[29-31] Both found that LTD at these synapses is greatly diminished. Interestingly, these mice show severe cerebellar ataxia, suggesting that this synaptic plasticity phenomenon is involved in the control of movement.

In the hippocampus, LTP can be obtained at the perforant path-dendate gyrus, mossy fiber-CA_3, Schaffer collateral-CA_1 synapses. It implicates NMDA receptor activation in CA_1 but not in CA_3 synapses. Activation of mGluRs was reported to facilitate the induction of LTP at CA_1 synapses.[117] Using MCPG, Collingridge and colleagues reported that activation of mGluRs was indeed necessary to induce LTP in CA_1 and CA_3[118], a result not reproduced by others.[119,120] In another approach, two groups examined the role of mGluRs in LTP using $mGluR_1$ knock-out mice. One group found that LTP can still be induced in CA_1, but is greatly diminished in CA_3,[31] suggesting that $mGluR_5$ rather than $mGluR_1$ may be involved in the CA_1 LTP. In agreement with this hypothesis, $mGluR_5$ but not $mGluR_1$ mRNA and immunoreactivity can easily be detected in CA_1 pyramidal neurons of wild-type animals (see ref. 62). However, Aiba and colleagues found that LTP is reduced in the hippocampal CA_1 area of $mGluR_1$ knock-out mice.[29] Impairment of spatial memory was observed after MCPG injection in the hippocampus,[121] and in $mGluR_1$ knock-out mice,[29-31] reinforcing the idea that mGluRs may play a role in spatial memory formation. In addition an impairment of associative memory was also observed in $mGluR_1$ knock-out mice generated by Aiba et al[29] (Table 5.3).

5.6 ROLE OF mGluRs IN NEURONAL SURVIVAL AND DEATH

The neurotoxic effect of Glu following ischemia, hypoglycemia and anoxia is mainly mediated via ionotropic receptors,[122] and therefore any modulatory action of mGluRs on ionotropic receptors may influence neuronal death. In general, the group I mGluRs potentiate NMDA and AMPA currents (for review see ref. 27). Therefore, facilitation of neuronal NMDA and AMPA induced neuronal death by group I mGluR agonists is expected. Accordingly, facilitation of the NMDA but not AMPA-mediated neurotoxicity by subtoxic doses of ACPD has been observed.[123,124] Moreover, injection of higher doses of 1S,3R-ACPD in the hippocampus results in the induction of limbic seizures and a loss of dentate granule cells, as well as of hippocampal CA_1 and CA_2 pyramidal cells.[125] Systemic administration of high doses of 1S,3R-ACPD in neonatal rats also induces convulsion and brain damage.[126] These were not inhibited by NMDA and AMPA receptor antagonists, indicating that activation of mGluRs may be, per se, sufficient to kill neurons. Note that in all these studies, the mGluR involved was not pharmacologically characterized. Curiously, a mGluR stimulated by ACPD and low doses of Quis (group I ?) attenuates Glu and kainate induced toxicity in cerebellar granule cells.[127]

Recently, a series of reports has described neuroprotective effects of mGluRs against various agents both in vivo[128,129] and *in vitro* (Table 5.4). The receptors involved were not always characterized but when they were, they belonged to group II or III.[63,130-135] In neuronal cultures, the neuroprotective effect of DCG-IV on short-term NMDA

Table 5.4. Survival and neuroprotective effects of mGluRs

	Toxicity induced by	**Attenuated or reversed by**	**mGluR implicated**	**References**
In Vivo				
Intraocular injections	NMDA	ACPD	?	Siliprandi et al. 1992 (128)
Focal ischemia		ACPD	?	Chiamulera et al. 1992 (129)
In Vitro				
Cortical cultures (neurons + astrocytes)	NMDA (500μM; 5 min)	ACPD	?	Koh et al. 1991 (130)
Cortical cultures (neurons + astrocytes)	NMDA (100μM; 10 min) AMPA Kainate	DCG - IV L - CCG I	Group II	Bruno et al. 1994 (134-135)
Cortical cultures (neurons + astrocytes)	NMDA (200μM; 5 min)	DCG - IV	Group II	Buisson et al. 1995 (132)
Cortical cultures (neurons + astrocytes)	β - AP 25-35*	ACPD DCG - IV L - AP4 L - SOP	Group II Group III	Copani et al. 1995 (131)
Mesencephalic cultures (neurons)	NMDA (100μM; 10 min)	ACPD DHPG L - CCGI	?	Ambrosini et al. 1995 (63)
Cortical cultures (neurons + astrocytes)	NMDA oxygen deprivation	S-4-CPG	Group II	Buisson et al. 1995 (133)
Cerebellar granule cells	Survival, no toxicity induction	L - AP4	Group III	Graham et al. 1995 (136)
Cerebellar granule cells	Glu kainate	ACPD QUIS	Group I (?)	Pizzi et al. 1993 (127)

* Fragment 25-35 of β - amyloid peptide

induced injury was confirmed by three laboratories[63,132,134] whereas only one[134,135] reported a positive neuroprotective effect of this group II specific agonist on the slowly induced neuronal death by kainate and AMPA. Note also an interesting effect of L-AP_4, the specific group III agonist on cerebellar granule cells survival, an effect similar to the well known survival actions of high K^+ concentrations or NMDA.[136] In conclusion, specific ligands for the different groups of mGluRs should help to discriminate the neuroprotective from the neurotoxic effects of mGluRs.

5.7 CONCLUSION AND PROSPECTS

Pharmacological tools and cloning have revealed a new family of receptors: the mGluRs. The effector and agonist recognition domains of these proteins have been identified, but the mechanisms by which the ligand induces the conformational changes required for G-protein activation remain to be determined. Some physiological roles for mGluRs have been identified, and the use of the gene knock-out techniques will probably reveal new important roles for each mGluR subtype, both at the cellular and behavioral levels. However, it is important to encourage development of more specific drugs to dissect the roles of mGluRs at multiple synapses, an essential step towards understanding the implication of glutamate synapses in brain function and disorder. Among the disorders that could benefit of mGluRs specific agonists and antagonists of the different subtypes one can tentatively identify ischemia and other degenerative disorders, motor and memory disorders as well as pain therapy.

References

1. Nakanishi S, Masu M. Molecular diversity and functions of glutamate receptors. Annu Rev Biophys Biomol Struc 1994; 23:319-348.
2. Nakanishi S. Molecular diversity of glutamate receptors and implications for brain function. Science 1992; 258:597-603.
3. Monaghan DT, Bridges RJ, Cotman CW. The excitatory amino acid receptors: their classes, pharmacology and distinct properties in the function of the central nervous system. Annu Rev Pharmacol Toxicol 1989; 29:365-402.
4. Hollmann M, Heinemann S. Cloned glutamate receptors. Annu Rev Neurosci 1994; 17:31-108.
5. Bettler B, Mulle C. AMPA and kainate receptors. Neuropharmacology 1995; 34:123-139.
6. Sladeczek F, Pin J-P, Récasens M et al. Glutamate stimulates inositol phosphate formation in striatal neurons. Nature 1985; 317:717-719.
7. Nicoletti F, Meek JL, Iadarola MJ et al. Coupling of inositol phospholipid metabolism with excitatory amino acid recognition sites in rat hippocampus. J Neurochem 1986; 46:40-46.
8. Nicoletti F, Iadarola MJ, Wroblewski JT et al. Excitatory amino acid rec-

ognition sites coupled with inositol phospholipid metabolism: developmental changes and interaction with α1-adrenoreceptors. Proc Natl Acad Sci (USA) 1986; 83:1931-1935.
9. Nicoletti F, Wroblewski JT, Novelli A et al. The activation of inositol phospholipid metabolism as a signal-transduction system for excitatory amino acids in primary cultures of cerebellar granule cells. J Neurosci 1986; 6:1905-1911.
10. Pearce B, Albrecht J, Morrow C et al. Astrocyte glutamate receptor activation promotes inositol phospholipid turnover and calcium flux. Neurosci Lett 1986; 72:335-340.
11. Sugiyama H, Ito I, Hirono C. A new type of glutamate receptor linked to inositol phospholipid metabolism. Nature 1987; 325:531-533.
12. Récasens M, Guiramand J, Nourigat A et al. A new quisqualate receptor subtype (sAA2) responsible for the glutamate-induced inositol phosphate formation in rat brain synaptoneurosomes. Neurochem Int 1988; 13:463-467.
13. Foster AC, Fagg GE. Acidic amino acid binding sites in mammalian neuronal membranes: their characteristics and relationship to synaptic receptors. Brain Res Rev 1984; 7:103-164.
14. Nawy S, Jahr CE. Suppression by glutamate of cGMP-activated conductance in retina bipolar cells. Nature 1990; 346:269-271.
15. Houamed KM, Kuijper JL, Gilbert TL et al. Cloning, expression, and gene structure of a G-protein-coupled glutamate receptor from rat brain. Science 1991; 252: 1318-1321.
16. Masu M, Tanabe Y, Tsuchida K et al. Sequence and expression of a metabotropic glutamate receptor. Nature 1991; 349:760-765.
17. Tanabe Y, Masu M, Ishii T et al. A family of metabotropic glutamate receptors. Neuron 1992; 8:169-179.
18. Abe T, Sugihara H, Nawa H et al. Molecular characterization of a novel metabotropic glutamate receptor $mGluR_5$ coupled to inositol phosphate/Ca^{2+} signal transduction. J Biol Chem 1992; 267:13361-13368.
19. Nakajima Y, Iwakabe H, Akazawa C et al. Molecular characterization of a novel retinal metabotropic glutamate receptor $mGluR_6$ with a high agonist selectivity for L-2-amino-4-phosphonobutyrate. J Biol Chem 1993; 268:11868-11873.
20. Saugstad JA, Kinzie JM, Mulvihill ER et al. Cloning and expression of a new member of the L-2-amino-4-phosphonobutyric acid-class of metabotropic glutamate receptors. Mol Pharmacol 1994; 45:367-372.
21. Okamoto N, Hori S, Akazawa C et al. Molecular characterization of a new metabotropic glutamate receptor $mGluR_7$ coupled to inhibitory cyclic AMP signal transduction. J Biol Chem 1994; 269:1231-1236.
22. Pin J-P, Waeber C, Prézeau L et al. Alternative splicing generates metabotropic glutamate receptors inducing different patterns of calcium release in *Xenopus* oocytes. Proc Natl Acad Sci USA 1992; 89:10331-10335.
23. Joly C, Gomeza J, Brabet I et al. Molecular, functional and pharmaco-

logical characterization of the metabotropic glutamate receptor type 5 splice variants: comparison with $mGluR_1$. J Neurosci 1995; 15:3970-3981.
24. Minakami R, Katsuki F, Sugiyama H. A variant of metabotropic glutamate receptor subtype 5: an evolutionally conserved insertion with no termination codon. Biochem Biophys Res Com 1993; 194:622-627.
25. Simoncini L, Haldeman BA, Yamagiwa T et al. Functional characterization of metabotropic glutamate receptor subtypes. Biophys J 1993; 64:A84.
26. Iversen L, Mulvihill E, Haldeman B et al. Changes in metabotropic glutamate receptor mRNA levels following global ischemia: Increase of a putative presynaptic subtype ($mGluR_4$) in highly vulnerable rat brain areas. J Neurochem 1994; 63:625-633.
27. Pin J-P, Duvoisin R. The metabotropic glutamate receptors: Structure and functions. Neuropharmacology 1995; 34:1-26.
28. Pin J-P, Bockaert J. Get receptive to metabotropic glutamate receptors. Curr Op Neur 1995; 5:342-349.
29. Aiba A, Chen C, Herrup K et al. Reduced Hippocampal long-term potentiation and context-specific deficit in associative learning in $mGluR_1$ mutant mice. Cell 1994; 79:365-375.
30. Aiba A, Kano M, Chen C et al. Deficient cerebellar long-term depression and impaired motor learning in $mGluR_1$ mutant mice. Cell 1994; 79:377-388.
31. Conquet F, Bashir ZI, Davies CH et al. Motor deficit and impairment of synaptic plasticity in mice lacking $mGluR_1$. Nature 1994; 372:237-243.
32. Masu M, Iwakabe H, Tagawa Y et al. Specific deficit of the On response in visual transmission by targeted disruption of the mGluRG gene. Cell 1995; 80:757-765.
33. Nakanishi S. Metabotropic glutamate receptors: synaptic transmission, modulation, and plasticity. Neuron 1994; 13:1031-1037.
34. Nakanishi S. Second-order neurons and receptor mechanisms in visual- and olfactory-information processing. Trends in Pharmacol. 1995; 18:359-364.
35. Brown EM, Gamba G, Riccardi D et al. Cloning and characterization of an extracellular Ca^{2+}-sensing receptor from bovine parathyroid. Nature 1993; 366:575-580.
36. Bockaert J. Les recepteurs à sept domaines transmembranaires: physiologie et pathologie de la transduction. Médecine/Science 1995; 11:382-394.
37. Savarese TM, Fraser CM. *In vitro* mutagenesis and the search for structure-function relationships among G protein-coupled receptors. Biochem J 1992; 283:1-19.
38. Ostrowski J, Kjelsberg MA, Caron MG et al. Mutagenesis of the b2-adrenergic receptor: How structure elucidates function. Annu Rev Pharmacol Toxicol 1992; 32:167-183.
39. Bockaert J. G-proteins and G-protein-coupled receptors: structure, function and interactions. Current Opinion in Neurobiology 1991; 1:32-42.
40. Fong TM, Huang RRC, Strader CD. The extracellular domain of the neurokinin-1 receptor is required for high-affinity binding of peptides. Biochemistry 1992; 31:11806-11811.

41. Cao Y, Gimpi G, Fahrenholz F. The amino-terminal fragment of the adenylate cyclase activating polypeptide (PACAP) receptor functions as a high affinity PACAP binding domain. Biochem Biophys Res Commun 1995; 212:673-680.
42. Stroop SD, Kuester RE, Serwold TF et al. Chimeric human calcitonin and glucagon receptors reveal two dissociable calcitonin interaction sites. Biochemisry 1995; 34:1050-1057.
43. O'Hara PJ, Sheppard PO, Thøgersen H et al. The ligand-binding domain in metabotropic glutamate receptors is related to bacterial periplasmic binding proteins. Neuron 1993; 11:41-52.
44. Takahashi K, Tsuchida K, Tanabe Y et al. Role of the large extracellular domain of metabotropic glutamate receptors in agonist selectivity determination. J Biol Chem 1993; 268:19341-19345.
45. Pollak MR, Brown EM, Chou Y-HW et al. Mutations in the human Ca^{2+}-sensing receptor gene cause familial hypocalciuric hypercalcemia and neonatal severe hyperparathyroidism. Cell 1993; 75:1297-1303.
46. Pollak MR, Brown EM, Estep HL et al. Autosomal dominant hypocalcaemia caused by a Ca^{2+}-sensing receptor gene mutation. Nature Genetics 1994; 8:303-307.
47. Stern-Bach Y, Bettler B, Hartley M et al. Agonist selectivity of glutamate receptors is specified by two domains structurally related to bacterial amino acid-binding proteins. Neuron 1994; 13:1345-1357.
48. Bennett JA, Dingledine R. Topology profile for a glutamate receptor: three transmembrane domains and a channel-lining reentrant membrane loop. Neuron 1995; 14:373-384.
49. Hollmann M, Maron C, Heinemann S. N-glycosylation site tagging suggests a three transmembrane domain topology for the glutamate receptor $GluR_1$. Neuron 1994; 13:1331-1343.
50. Spengler D, Waeber C, Pantaloni C et al. Differential signal transduction by five splice variants of the PACAP receptor. Nature 1993; 365:170-175.
51. Pin J-P, Joly C, Heinemann SF et al. Domains involved in the specificity of G protein activation in phospholipase C coupled metabotropic glutamate receptor. EMBO J. 1994; 13:342-348.
52. Gomeza J, Joly C, Kühn R et al. The second intracellular loop of mGluR1 cooperates with the other intracellular domains to control coupling to G-protein. J. Biol. Chem. 1995; 271:2199-2205.
53. Brabet I, Mary S, Bockaert J et al. Phenylglycine derivatives discriminate between mGluR1 and mGluR5 mediated responses. Neuropharmacology. 1995; 34:895-903.
54. Ito I, Kohda A, Tanabe S et al. 3,5-dihydroxyphenyl-glycine: a potent agonist of metabotropic glutamate receptors. 1992; 3:1013-1016.
55. Schoepp DD, Goldsworthy J, Johnson BG et al. 3,5-Dihydroxyphenylglycine is a highly selective agonist for phosphoinositide-linked metabotropic glutamate receptors in the rat hippocampus. J Neurochem 1994; 63:769-772.
56. Gereau RW,Conn PJ. Roles of specific metabotropic glutamate receptor

subtypes in regulation of hippocampal CA1 pyramidal excitability. J Neurophysiol 1995; 74:122-129.

57. Manzoni O, Bockaert J. Metabotropic glutamate receptors inhibiting excitatory synapses in the CA_1 area of the rat hippocampus. Eur J Neurosc 1995; 7: 2518-2523.
58. Hayashi Y, Tanabe Y, Aramori I et al. Agonist analysis of 2-(carboxycyclopropyl) glycine isomers for cloned metabotropic glutamate receptor subtypes expressed in Chinese hamster ovary cells. Br J Pharmacol 1992; 107:539-543.
59. Hayashi Y, Momiyama A, Takahashi T et al. Role of a metabotropic glutamate receptor in synaptic modulation in the accessory olfactory bulb. Nature 1993; 366:687-690.
60. Aramori I, Nakanishi S. Signal transduction and pharmacological characteristics of a metabotropic glutamate receptor, $mGluR_1$, in transfected CHO cells. Neuron 1992; 8:757-765.
61. Saugstad JA, Segerson T, Westbrook GL. L-2-amino-3-phosphonopropionic acid competitively antagonizes metabotropic glutamate receptors 1a and 5 in *Xenopus* oocytes. Eur J Pharmacol 1995:395-397.
62. Bockaert J, Gomeza J, Brabet I et al. $mGluR_1$ and $mGluR_5$ glutamate receptors: molecular biology, pharmacology, roles in hippocampal synaptic transmission. In: Eds. Long-Term Potentiation - A Debate of Current Issues. MIT Press, 1995: in press.
63. Ambrosini A, Bresciani L, Fracchia S et al. Metabotropic glutamate receptors negatively coupled to adenylate cyclase inhibit N-Methyl-D-aspartate receptor activity and prevent neurotoxicity in mesencephalic neurons. Mol Pharmacol 1995; 47:1057-1063.
64. Hayashi Y, Sekiyama N, Nakanishi S et al. Analysis of agonist and antagonist activities of phenylglycine derivatives for different cloned metabotropic glutamate receptor subtypes. J Neurosci 1994; 14:3370-3377.
65. Tanabe Y, Nomura A, Masu M et al. Signal transduction, pharmacological properties, and expression patterns of two rat metabotropic glutamate receptors, $mGluR_3$ and $mGluR_4$. J Neurosci 1993; 13:1372-1378.
66. Jane DE, Jones PLSJ, Pook PC-K et al. Actions of two new antagonists showing selectivity for different sub-types of metabotropic glutamate receptor in the neonatal rat spinal cord. Br J Pharmacol 1994; 112:809-816.
67. Manzoni O, Castillo P, Nicoll RA. Pharmacology of metabotropic glutamate receptors at the mossy fiber synapses of the guinea pig hippocampus. Neuropharmacology 1995; 34:965-971.
68. Watkins J,Collingridge G. Phenylglycine derivatives as antagonists of metabotropic glutamate receptors. Trends Neurosci 1994; 15:333-342.
69. Pickering DS, Thomsen C, Suzdak PD et al. A comparison of two alternatively spliced forms of a metabotropic glutamate receptor coupled to phosphoinositides turnover. J Neurochem 1993; 61:85-92.
70. Thomsen C, Mulvihill ER, Haldeman B et al. A pharmacological characterization of the mGluR1α subtype of the metabotropic glutamate

receptor expressed in a cloned baby hamster kidney cell line. Brain Res 1993; 619:22-28.
71. Prézeau L, Carette J, Helpap B et al. Pharmacological characterization of metabotropic glutamate receptors in several types of brain cells in primary cultures. Mol Pharmacol 1994; 45:570-577.
72. Dumuis A, Pin JP, Oomagari K et al. Arachidonic acid released from striatal neurons by joint stimulation of ionotropic and metabotropic quisqualate receptors. Nature 1990; 347:182-184.
73. Dumuis A, Sebben M, Fagni L et al. Stimulation of arachidonic acid release by glutamate receptors depends on Na^+/Ca^{2+} exchanger in neuronal cells. Mol Pharmacol 1993; 43:976-981.
74. Ikeda SR, Lovinger DM, McCool BA et al. Heterologous expression of metabotropic glutamate receptor in adult rat sympathetic neurons: subtype-specific coupling to ion channels. Neuron 1995; 14:1029-1036.
75. Chavis P, Fagni L, Bockaert J et al. Modulation of calcium channels by metabotropic glutamate receptors in cerebellar granule cells. Neuropharmacology 1995; 34:929-937.
76. Chavis P, Nooney JM, Bockaert J et al. Facilitatory coupling between a glutamate metabotropic receptor and dihydropyridine-sensitive calcium channels in cultured cerebellar granule cells. J Neurosci 1995; 15:135-143.
77. Trombley PQ, Westbrook GL. L-AP_4 inhibits calcium currents and synaptic transmission via a G-protein-coupled glutamate receptor. J Neurosci 1992; 12:2043-2050.
78. Lester RAJ, Jahr CE. Quisqualate receptor-mediated depression of calcium currents in hippocampal neurons. Neuron 1990; 4:741-749.
79. Sayer RJ, Schwindt PC, Crill WE. Metabotropic glutamate receptor-mediated suppression of L-type calcium current in acutely isolated neocortical neurons. J Neurophysiol 1992; 68:833-842.
80. Swartz KJ, Merrit A, Bean BP et al. Protein kinase C modulates glutamate receptor inhibition of Ca^{2+} channels and synaptic transmission. Nature 1993; 361:165-168.
81. Swartz KJ, Bean BP. Inhibition of calcium channels in rat CA_3 pyramidal neurons by a metabotropic glutamate receptor. J Neurosci 1992; 12:4358-4371.
82. Baskys A, Bernstein NK, Barolet AW et al. NMDA and quisqualate reduce a Ca^{2+}-dependent K^+ current by a protein kinase-mediated mechanism. Neurosci Lett 1990; 112:76-81.
83. Charpak S, Gähwiler BH, Do KQ et al. Potassium conductances in hippocampal neurons blocked by excitatory amino-acid transmitters. Nature 1990; 347:765-767.
84. Gerber U, Sim JA, Gähwiler BH. Reduction of potassium conductances mediated by metabotropic glutamate receptors in rat CA_3 pyramidal cells does not require protein kinase C or protein kinase A. Eur J Neurosci 1992; 4:792-797.
85. Glaum SR, Miller RJ. Metabotropic glutamate receptors mediate excitatory transmission in the nucleus of the solitary tract. J Neurosci 1992; 12:2251-2258.

86. Fagni L, Bossu J-L, Bockaert J. Activation of a large-conductance Ca^{2+}-dependent K^+-channel by stimulation of glutamate phosphoinositide-coupled receptors in cultured cerebellar granule cells. Eur J Neurosci 1991; 3:778-789.
87. Prézeau L, Manzoni O, Homburger V et al. Characterization of a metabotropic glutamate receptor: direct negative coupling to adenylate cyclase and involvement of a pertussis toxin-sensitive G protein. Proc Natl Acad Sci USA 1992; 89:8040-8044.
88. Chavis P, Shinozaki H, Bockaert J et al. The metabotropic glutamate receptor types 2/3 inhibit L-type calcium channels via a Pertussis toxin-sensitive G-protein in cultured cerebellar granule cells. J Neurosci 1994; 14:7067-7076.
89. Casabona G, Genazzani AA, DiStefano M et al. Developmental changes in the modulation of cyclic AMP formation by the metabotropic glutamate receptor agonist 1S,3R-aminocyclopentane-1,3-dicarboxylic acid in brain slices. J Neurochem 1992; 59:1161-1163.
90. Schoepp DD, Johnson BG. Metabotropic glutamate receptor modulation of cAMP accumulation in the neonatal rat hippocampus. Neuropharmacology 1993; 32:1359-1365.
91. Winder DG, Conn PJ. Activation of metabotropic glutamate receptors in the hippocampus increases cyclic AMP accumulation. J Neurochem 1992; 59:375-378.
92. Winder DG, Conn PJ. Activation of metabotropic glutamate receptors increases cAMP accumulation in hippocampus by potentiating responses to endogenous adenosine. J Neurosci 1993; 13:38-44.
93. Winder DG, Conn PJ. Metabotropic glutamate receptor (mGluR)-mediated potentiation of cyclic AMP responses does not require phosphoinositide hydrolysis: mediation by group II-like mGluR. J Neurochem 1995; 64:592-599.
94. Tang WJ, Gilman A. Type-specific regulation of adenylyl cyclase by G protein βγ subunits. Science 1991; 254:1500-1503.
95. Saugstad JA, Kinzie JM, Mulvihill ER et al. Cloning an expression of a new member of the L-2-amino-4-phosphonobutyric acid-class of metabotropic glutamate receptors. Mol Pharmacol 1994; 45:367-372.
96. Thomsen C, Kristensen P, Mulvihill E et al. L-2-amino-4-phosphonobutyrate (L-AP_4) is an agonist at the type IV metabotropic glutamate receptor which is negatively coupled to adenylate cyclase. Eur J Pharmacol 1992; 227:361-362.
97. Cartmell J, Kemp JA, Alexander SPH et al. Endogenous adenosine regulates the apparent efficacy of 1-aminocyclopentyl-1S,3R-dicarboxylate inhibition of forskolin-stimulated cyclic AMP accumulation in rat cerebral cortical slices. J Neurochem 1993; 60:780-782.
98. Cartmell J, Kemp JA, Alexander SPH et al. Modulation of cyclic AMP formation by putative metabotropic receptor agonists. Br J Pharmacol 1994; 111:364-369.
99. Kemp M, Roberts P, Pook P et al. Antagonism of presynaptically mediated depressant responses and cyclic AMP-coupled metabotropic

glutamate receptors. Eur J Pharmacol (Mol Pharmac Sec) 1994; 266:187-192.
100. Nomura A, Shigemoto R, Nakamura Y et al. Developmentally regulated postsynaptic localization of a metabotropic glutamate receptor in rat rod bipolar cells. Cell 1994; 77:361-369.
101. Shiells RA, Falk G. Glutamate receptors of rod bipolar cells are linked to a cyclic GMP cascade via a G-Protein. Proc R Soc Lond [Biol] 1990; 242:91-94.
102. Schoepp DD, Bockaert J, Sladeczek F. Pharmacological and functional characteristics of metabotropic excitatory amino acid receptors. Trends Pharmac Sci 1990; 11:508-515.
103. Birse EF, Eaton SA, Jane DE et al. Phenylglycine derivatives as new pharmacological tools for investigating the role of metabotropic glutamate receptors in the central nervous system. Neurosci 1993; 52:481-488.
104. Boss V, Conn PJ. Metabotropic excitatory amino acid receptor activation stimulates phospholipase D in hippocampal slices. J Neurochem 1992; 59:2340-2343.
105. Boss V, Nutt KM, Conn PJ. L-cystein sulfinic acid as an endogenous agonist of a novel metabotropic receptor coupled to stimulation of phospholipase D activity. Mol Pharmacol 1994; 45:1177-1182.
106. Stella N, Tence M, Glowinski J et al. Glutamate-evoked release of arachidonic acid from mouse brain astrocytes. J Neurosci 1994; 14:568-575.
107. Guérineau NC, Gähwiler BH, Gerber U. Reduction of resting K^+ current by metabotropic glutamate and muscarinic receptors in rat CA_3 cells–mediation by G-proteins. J Physiol (Lond) 1994; 474:27-33.
108. Glaum SR, Slater NT, Rossi DJ et al. Role of metabotropic glutamate (ACPD) receptors at the parallel fiber-Purkinje cell synapse. J Neurophysiol 1992; 68:1453-1462.
109. Mc Cormick DA, Krosigk MV. Corticothalamic activation modulates thalamic firing through glutalate "metabotropic" receptors. Proc Natl Acad Sci (USA) 1992; 89:2774-2778.
110. Guerineau NC, Bossu JL, Gähwiler BH et al. Activation of a non selective cationic conductance by metabotropic glutamatergic and muscarinic agonists in CA_3 pyramidal neurons of rat hippocampus. J Neurosci 1995; 15:4395-4407.
111. Chabre M, Deterre P. Molecular mechanism of visual transduction. Eur J Biochem 1989; 179:255-256.
112. Bachelor AM, Madge DJ, Garthwaite J. Synaptic activation of metabotropic glutamate receptors in the parallel fibre-Purkinje cell pathway in rat cerebral slices. Neuropharmacology 1994; 32:11-20.
113. Eaton SA, Birse EF, Wharton B et al. Mediation of thalamic sensory responses *In vivo* by ACPD-activated excitatory amino acid receptors. Eur J Neurosci 1993; 5:186-189.
114. Kaba H, Hayashi Y, Higuchi T et al. Induction of an olfactory memory by the activation of a metabotropic glutamate receptor. Science 1994; 265: 262-264.

115. Young MR, Fleetwood-Walker SM, Mitchell R et al. Evidence for a role of metabotropic glutamate receptors in sustained nociceptive inputs to dorsal horn neurons. Neuropharmacology 1994; 33:141-144.
116. Shigemoto R, Abe T, Nomura S et al. Antibodies inactivating $mGluR_1$ metabotropic glutamate receptor block long-term depression in cultured Purkinje cells. Neuron 1994; 12:1245-1255.
117. Aniksztejn L, Otani S,Ben-Ari Y. Quisqualate metabotropic receptors modulate NMDA currents and facilitate induction of long-term potentiation through protein kinase C. Eur J Neurosci 1992; 4:500-505.
118. Bashir ZI, Bortolotto ZA, Davies CH et al. Induction of LTP in the hippocampus needs synaptic activation of glutamate metabotropic receptors. Nature 1993; 363:347-350.
119. Chinestra P, Aniksztejn L, Diabira D et al. (RS)-α-Methyl-4-carboxyphenylglycine neither prevents induction of LTP nor antagonizes metabotropic glutamate receptors in CA1 hippocampal neurons. J Neurophysiol 1993; 70:2684-2689.
120. Manzoni OJ, Weisskop MG, Nicoll RA. MCPG antagonizes metabotropic glutamate receptors but not long-term potentiation in the hippocampus. Eur J Neurosci 1994; 6:1050-1054.
121. Riedel G, Wetzel W, Reymann KG. (R,S)-alpha-methyl-4-carboxyphenylglycine (MCPG) blocks spatial learning in rats and long-term potentiation in the dentate gyrus *in vivo*. Neurosci Lett 1994; 167:141-144.
122. Choi DW. Glutamate neurotoxicity and diseases of the nervous system. Neuron 1988; 1:623-634.
123. McDonald JW, Schoepp DD. The metabotropic excitatory amino acid receptor agonist 1S,3R-ACPD selectively potentiates N-methyl-D-aspartate-induced brain injury. Eur J Pharmacol 1992; 215:353-354.
124. Bruno V, Copani A, Knopfel T et al. Activation of metabotropic glutamate receptors coupled to inositol phosphoplipid hydrolysis amplifies NMDA-induced neuronal degeneration in cultured cortical cells. Neuropharmacology 1995; 34:1089-1098.
125. Sacaan AI, Schoepp DD. Activation of hippocampal metabotropic excitatory amino acid receptors leads to seizures and neuronal damage. Neurosci Lett 1992; 139:77-82.
126. McDonald JW, Fix AS, Tizzano JP et al. Seizures and brain injury in neonatal rats induced by 1S,3R-ACPD, a metabotropic glutamate receptor agonist. J Neurosci 1993; 13:4445-4455.
127. Pizzi M, Fallacara C, Arrighi V et al. Attenuation of excitatory amino acid toxicity by metabotropic glutamate receptor agonists and aniracetam in primary cultures of cerebellar granule cells. J Neurochem 1993; 61:683-689.
128. Siliprandi R, Lipartiti M, Fadda E et al. Activation of the glutamate metabotropic receptor protects retina against N-methyl-D-aspartate toxicity. Eur J Pharmacol 1992; 219:173-174.
129. Chiamulera C, Albertini P, Valerio E et al. Activation of metabotropic receptors shows a neuroprotective effect in a rodent model of focal ischemia. Eur J Pharmacol 1992; 216:335-336.

130. Koh J-Y, Palmer E, Cotman CW. Activation of the metabotropic glutamate receptor attenuates N-methyl-D-aspartate neurotoxicity in cortical cultures. Proc Natl Acad Sci USA 1991; 88:9431-9435.
131. Copani A, Bruno V, Battaglia G et al. Activation of metabotropic glutamate receptors protects cultured neurons against apoptosis induced by β-amyloid peptide. Mol Pharmacol 1995; 47:890-897.
132. Buisson A, Yu S, Choi D. DCG-IV selectively attenuates rapidly-triggered NMDA-induced neurotoxicity on cortical neurons. Eur J Neurosci 1995; in press.
133. Buisson A, Choi DW. The inhibitory mGluR agonist, S-4-carboxy-3-hydroxy-phenylglycine selectively attenuates NMDA neurotoxicity and oxygen-glucose deprivation-induced neuronal death. Neuropharmacology 1995; 34:1081-1087.
134. Bruno V, Copani A, Battaglia G et al. Protective effect of the metabotropic glutamate receptor agonist, DCG-IV, against excitotoxic neuronal death. Eur J Pharmacol 1994; 256:109-112.
135. Bruno V, Battaglia G, Copani A et al. Activation of class II or III metabotropic glutamate receptors protects cultured cortical neurons against excitotoxic degeneration. Eur J Neurosci 1995; 7:in press.
136. Graham ME, Burgoyne RD. Activation of metabotropic glutamate receptors by L-AP_4 stimulates survival of rat cerebellar granule cells in culture. Eur J Pharmacol 1994; 288:115-123.

CHAPTER 6

The 5-HT4, 5-HT5, 5-HT6 and 5-HT7 Receptors: New Receptors and New Tools to Study Them

Ronald S. Oosting, Jose J. Lucas, Regis Grailhe, Mark J.S. Heath, and René Hen

6.1 INTRODUCTION

Serotonin (5-hydroxytryptamine, 5-HT) is a major neurotransmitter and plays a modulatory role in a number of brain functions. For instance, serotonin is involved in sleep, appetite, pain perception, locomotion, thermoregulation and sexual activity.[1] Serotonergic drugs are used in the treatment of a number of pathological states such as migraine, depression, and anxiety, and a serotonin deficit has been associated with behaviors such as suicide, impulsive violence, depression, and alcoholism.[2]

The various actions of 5-HT are mediated by a large number of receptors. Until now, 14 different mammalian 5-HT receptor subtypes have been cloned. With the exception of the 5-HT3 receptor, which is a ligand-gated ion channel, the 5-HT receptors belong to the large family of G-protein coupled seven transmembrane (7-TM) receptors. Sequence comparisons reveal striking amino acid conservation, particularly within transmembrane domains (see Fig. 6.1). The classification of the various 5-HT receptors takes into account: a) their primary structure, b) their pharmacological profile, and c) their coupling to the various effector systems.[3-5] The 5-HT 7-TM receptors may either be coupled to adenylate cyclase (negatively: 5-HT1A, 5-HT1B, 5-HT1D, 5-HT1E, and 5-HT1F and positively: 5-HT4, 5-HT6, and 5HT7) or to

Recombinant Cell Surface Receptors: Focal Point for Therapeutic Intervention, edited by Michael J. Browne.

phospholipase C (5-HT2A, 5-HT2B, and 5-HT2C). So far the effector system for the 5-ht5A and 5-ht5B receptors has not been identified.

5-HT and its receptors are found throughout the animal kingdom. Serotonin receptors have been found in lower organisms such as worms. This finding suggests that the diversity of the 5-HT receptors, as present today, is the result of more than 700 million years of molecular evolution.[6]

In this chapter we will focus on the most recently cloned 5-ht receptors (5-HT4, 5-ht5A, 5-ht5B, 5-HT6 and 5-HT7). Key features are summarized in Tables 6.1 and 6.2. In addition, the potential contribution of gene/mRNA knockout experiments to the unraveling of their physiological and pathological function will be discussed.

6.2 THE 5-HT4 RECEPTOR

The rat 5-HT4 receptor gene has been cloned from rat brain RNA by reverse transcriptase-polymerase chain reaction (RT-PCR) cloning.[7] Two different cDNAs were isolated that presumably are the result of an alternative splicing event that generates a short (387 amino acids) and a long isoform (406 amino acids) diverging only in the C-terminal tail. Interestingly, the long isoform contains a protein kinase C phosphorylation site not present in the short isoform.

Table 6.1. Characteristics of the 5-HT4, 5-ht5, 5-HT6 and 5-HT7 receptors.

Receptor	Coupling	Chromosomal localization	Number of amino acids	Position of introns in coding sequence
5-HT4	adenylate cyclase (+)	?	387/406 * (rat) 390 (human)	C-terminal
5-ht5A	?	5A3-B (mouse) 7q35-36b (human)	357 (rat) 357 (mouse) 357 (human)	I3
5-ht5B	?	1E4-G (mouse) 2q11-13 (human)	370 (rat) 370 (mouse)	I3
5-HT6	adenylate cyclase (+)	?	436/437 (rat)	I3, E3
5-HT7	adenylate cyclase (+)	19D (mouse) 10q23.3-24.3 (human)	435/448* (rat) 448 (mouse) 445 (human) 446 (guinea pig) 426 (xenopus)	I2, C2

* The result of alternative splicing. Data are taken from ref. 5. I : intracellular loop; E: extracellular loop.

When expressed in mammalian cells, both isoforms of the 5-HT4 receptor displayed a high affinity for [3H]-GR113808.[7] The displacement profile (cisapride, > 5-HT > renzapride > 5-MeOT > zacopride > α-Me-5-ht > 5-CT) was very similar for both isoforms and correlated well with that of the 5-HT4 receptor as defined by functional and radioligand binding assays.[8,9]

In transiently transfected COS-7 cells, stimulation of the 5-HT4

Table 6.2. Regional distribution of 5-ht5A, 5-ht5B, 5-HT6 and 5-HT7 receptor mRNA's

brain area		5-ht5A (mouse)	5-ht5B (mouse)	5-HT6 (rat)	5-HT7 (mouse)
cerebral cortex	cingulate cortex	++		++	+++
	frontal cortex	++		++	++
	parietal cortex	++		++	++
	entorhinal cortex	++		++	+++
basal ganglia	striatum			+++	
	accumbens nucleus			+++	++
	globus pallidus				+
septum	lateral septal nucleus			++	
	septohippocampal nucl.				+++
hippocampus	CAI pyramidal cell layer	++	+++	++	
	CA2	++		++	+++
	CA3	++		++	++
	dendate gyrus	++		++	
	tenia tecta				+++
	fimbria fornix				+++
hypothalamus	medial mammillary nucl.				++
	lateral mammillary nucl.				++
thalamus	paraventricular th. nucl.				++
	anteroventral th. nucl.				++
	anteromedial th. nucl.				++
	mediodorsal th. nucl.				++
	parafascicular th. nucl.				++
	ventral th. nucl.				+
habenula	medial habenula	++	+++		
	lateral habenula		++		
visual system	geniculate nucleus				++
	superior colliculus				+
olfactory system	olfactory bulb	+		++	
	olfactory tubercules			+++	
amygdala	amydalohippocamp. area				++
medula oblongata/pons	dorsal raphe nucleus		+		+
cerebellum	granular layer	++			

These data are taken from refs 20, 25, and 35 and our unpublished data.

receptor produced, as expected, an increase in cAMP levels,[7] but, in contrast to binding studies, the two isoforms behaved differently in pharmacology experiments. For instance, cisapride, renzapride, and zacopride were equipotent in activating adenylate cyclase in cells expressing the long isoform, but exhibited different potencies in cells transfected with the short isoform. Additionally, both forms responded differentially to the non hydrolyzable analog of GTP, GTPγS. While GTPγS had no effect on the short isoform, it induced a reduction in the affinity of 5-HT for the long isoform.

RT-PCR experiments performed on different rat brain structures have revealed the presence of the long isoform in striatum, brain stem, thalamus, hippocampus, and olfactory bulb, while the short form is detected only in the striatum.[7] In the periphery, the ileum, colon and bladder express both isoforms, while the heart atrium expresses only the short form. These data are compatible with the distribution of the protein deduced from autoradiographic studies with the 5-HT4 receptor selective radioligands [3H]-GR113808[10-12] and [125I]-SB204070.[13] High densities of 5-HT4 binding sites were found in olfactory tubercle, globus pallidus, striatum, substantia nigra, and hippocampus of the guinea pig,[10] rat,[10] and human brain.[11,12]

The generation by alternative splicing of receptors differing only in their C-terminal tail is a characteristic shared also by the somatostatin-SSTR2,[14] the prostaglandin EP3[15] and the pituitary adenylate cyclase activating polypeptide (PACAP)[16] receptors. Such alternative splicing leads in the case of the EP3 and PACAP receptors to coupling with distinct G proteins.[15,16] The possibility that each isoform of the 5-HT4 receptor displays a differential coupling to G proteins is supported by their different sensitivity to GTPγS.

6.3 THE 5-HT5A AND 5-HT5B RECEPTORS

The 5-ht5A receptor (357 amino acids) has been cloned from mouse,[17] rat,[18] and human,[19] whereas the 5-ht5B receptor (370 amino acids) has been cloned only from mouse[20] and rat.[18,21] The 5-ht5A and 5-ht5B receptors are very similar to one another (77% homology), but the homology to other 5-HT receptors is much less (See Fig. 6.1).

By contrast with all other cloned 5-HT receptors, no effect on adenylate cyclase or phospholipase C (PLC) activities could be detected when expressing the 5-ht5 receptors in COS-7 or NIH-3T3 cells.[17,18] In the case of the 5-ht5B receptor, however, a fraction of the binding sites observed in transfected COS-1 cells were displaced by GTP analogs, suggesting that this receptor couples to G proteins in these cells.[21] It is, therefore, possible that the 5-ht5 receptors interact with an effector system other than adenylate cyclase or PLC.

The 5-ht5A receptor is expressed exclusively in the CNS. In situ hybridization experiments have revealed the presence of 5-ht5A mRNAs in the cerebral cortex, hippocampus, granule cells of the cerebellum,

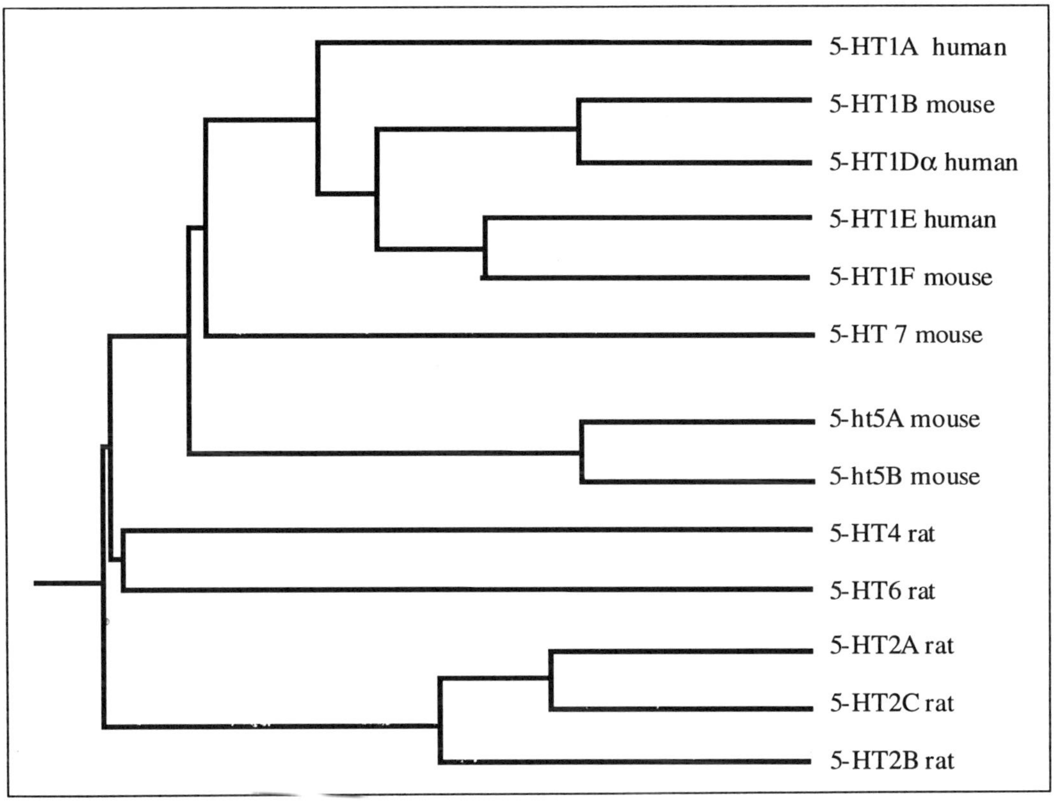

Fig 6.1. Dendrogram of the amino acid sequences of the cloned mammalian 7-TM 5-HT receptors. For comparison, the amino acid sequences corresponding to the N- and C-terminals and to the third cytoplasmic loop were excluded. The length of the horizontal lines are inversely proportional to the percentage of homology between receptors or groups of receptors.

medial habenula, amygdala, septum, several thalamic nuclei and tufted cells of the olfactory bulb[17,18] (see also Table 6.1).

The 5-ht5B receptor is expressed mainly in the limbic system and in situ hybridization experiments performed on mouse brain sections revealed the presence of 5-ht5B receptor mRNA predominantly in the CA1 field of the hippocampus, the habenula and the dorsal raphe nucleus[20] (see also Table 6.1). Low levels of expression were also found in the entorhinal and piriform cortex, subiculum and olfactory bulb in the rat.[18,21]

The mouse and the rat 5-ht5A receptors contain several potential phosphorylation sites for protein kinase C, cAMP dependent protein kinase and two potential sites of N-linked glycosylation. Analysis of genomic DNA fragments containing the mouse 5-ht5A gene indicated the presence of one intron about 8 kb pair long, located in the middle of the third cytoplasmic loop.[20] The gene encoding the 5-ht5A receptor is located on human chromosome 7 at position 7q36.[22] This locus also contains the holoprosencephaly type III mutation that results in abnormal brain development, raising the possibility that the 5-ht5A receptor plays a role in brain development.[18,22]

Both the 5-ht5 receptors have a high affinity for LSD and ergot derivatives and might therefore mediate some of the effects of these drugs.[23]

Until now the 5-ht5 receptors have not been identified in in vivo preparations: therefore in line with IUPHAR convention lower case letters are used.

6.4 THE 5-HT6 RECEPTOR

The 5-HT6 receptor has been cloned from rat and consists of a polypeptidic chain of 437[24] or 436 amino acids[25] and displays little homology with other 5-HT receptors. The two published amino acid sequences differ in their C terminal tail. However, the nucleotide sequences are identical except for one nucleotide which is absent from one of the sequences–resulting in a frame shift. The 5-HT6 receptor contains one potential site for N-linked glycosylation and several potential sites for phosphorylation by cAMP dependent protein kinase and protein kinase C. At least one intron is located in the middle of the third cytoplasmic loop.[24,25]

The most remarkable pharmacological feature of this receptor is its high affinity for typical and atypical antipsychotic drugs such as loxapine and clozapine as well as tricyclic antidepressant drugs (for example, amoxipine and clomiprimine).[24] Activation of the 5-HT6 receptor expressed in HEK-293 cells[24] or COS-7 cells[25] results in stimulation of adenylate cyclase.

In the brain, 5-HT6 mRNA is found in the caudate putamen, olfactory tubercle, and nucleus accumbens as well as in hippocampus (CA1, CA2 and CA3 fields, and dentate gyrus), olfactory bulb, and cerebral cortex.[24,25] In peripheral tissues a low expression was detected by Northern blot in rat stomach and guinea pig adrenals.[24,25]

The first evidence for in vivo existence of the 5-HT6 receptor has been recently reported.[26] In pig brain caudate membranes, 5-HT agonists induced a concentration dependent stimulation of adenylate cyclase activity with a rank order of potency (5-HT $\geq$ 5-methoxytryptamine > 5-CT > sumatriptan) very similar to that of the 5-HT6 receptor. Additional pharmacological similarities (including antagonism by clozapine) along with a common localization in striatum suggest that this receptor corresponds to the cloned 5-HT6 receptor.

In addition, the pharmacological properties of the 5-HT6 receptor are similar to those of 5-HT receptors positively coupled to adenylyl cyclase that were characterized in the NCB-20 neuroblastoma cell line[27] and in N18TG2 cells.[28]

The high affinity of the 5-HT6 receptor for atypical antipsychotic drugs such as clozapine led to the hypothesis that this receptor could mediate some of the effects of these drugs.[24,29] In particular, the lack of extrapyramidal side effects of clozapine could be the result of an action at striatal 5-HT6 receptors.[29] In a recent study, the affinities of over 30 typical and atypical antipsychotic drugs for the rat 5-HT6 and 5-HT7 receptors in transfected cells were measured.[29] To date, no consistent pattern has emerged as to whether the 5-HT6 and/or 5-HT7 receptors might contribute to the effects of these drugs. While some atypical antipsychotic drugs (for example olanzepine) have a higher affinity for the 5-HT6 receptor, others such as risperidone and tiospirone have a higher affinity for the 5-HT7 receptor.

6.5 THE 5-HT7 RECEPTOR

The 5-HT7 receptor has been cloned in human,[30] rat,[31-34] mouse,[35] guinea pig,[36] and *Xenopus laevis*,[37] and a putative homologue was found in *Drosophila melanogaster*.[38] Such evolutionary conservation might imply an important role for this receptor. The gene for the rat 5-HT7 receptor has been shown to contain at least two introns in its coding sequence,[33,34] one located after the third transmembrane domain, and the other one close to the end of the coding region. In the rat, Lovenberg et al[31] isolated a 435 amino acid cDNA, while Ruat et al[33] isolated a 448 amino acid cDNA. These cDNAs differ only in their C-terminus and probably result from an alternative splicing event. As mentioned for the 5HT4 receptor, the generation of receptors differing in their C-terminal may lead to a differential G protein coupling. When expressed in mammalian cells, both forms of the 5-HT7 receptor were positively coupled to adenylate cyclase.[31-37]

The 5-HT7 receptor is expressed mainly in the CNS. In situ hybridization experiments performed in rat and guinea pig, detected the 5-HT7 mRNA in hippocampus (pyramidal cells of CA2 and CA3 layers, tenia tecta and fimbria fornix), hypothalamus, thalamus, amygdaloid complex, cortex, superior colliculus and dorsal and paramedian raphe nuclei[38-43] (see also Table 6.1). In peripheral tissues, a low level of expression was detected by Northern or RT-PCR analysis in heart,[35] coronary arteries,[30] stomach,[33] ileum,[33] and spleen.[34]

The pharmacological profile of the cloned 5-HT7 receptor is similar to that of 5-HT receptors positively coupled to adenylate cyclase and found in the cardiovascular and gastrointestinal systems.[39-41]

In rat hypothalamus, a population of [3H]5-HT binding sites was recently identified that might also correspond to the 5-HT7 receptor.[42] Furthermore, these binding sites were downregulated after chronic treatment with fluoxetine,[42] suggesting that the 5-HT7 receptor might be involved in the therapeutic effects of anti-depressant drugs.

Due to the affinity of the 5-HT7 receptor for 8-OH-DPAT, this receptor might correspond to previously reported 5-HT1A-like receptors that are positively coupled to adenylate cyclase.[43-47] Such receptors have been suggested to play a role in circadian rhythms.[31,46] Several 5-HT agonists including 8-OH-DPAT are able to induce a phase shift in spontaneous neuronal activity in the suprachiasmatic nucleus (SCN).[31,46] Among the cloned 5-HT receptors, the best candidate to mediate this effect is the 5-HT7 receptor. In addition, a 5-HT7-like receptor has recently been shown to inhibit a $GABA_A$ activated current in cultured SCN neurones.[48] However, due to the limited pharmacological characterization of these phenomena and the fact that 5-HT7 transcripts could not be detected in SCN, it is not clear whether the 5-HT7 receptor is responsible for these effects.

Finally, as in the case of the 5-HT6 receptor, the high affinity of the 5-HT7 receptor for atypical neuroleptics such as risperidone and clozapine has led to the hypothesis that this receptor might also play a role in certain neuropsychiatric disorders.[29]

6.6 NEW WAYS TO INVESTIGATE (5-HT) RECEPTOR FUNCTION

The diversity of the 5-HT receptors and their overlapping distribution in the central nervous system make the analysis of their physiological and pathological roles extremely difficult. Selective agonists and antagonists have been developed for a number of 5-HT receptors. However, for some recently cloned receptors there are as yet no selective ligands.

The availability of the genes encoding these receptors has enabled the development of a genetic approach to study 5-HT receptor function. Instead of targeting a particular receptor protein, the "genetic pharmacology" approach targets the gene or the mRNA encoding this protein. Two approaches are presently used: 1) antisense knockout and 2) gene knockout.

6.6.1 Antisense Knockout

Antisense oligodeoxynucleotides inhibit gene expression by hybridization to coding (sense) sequences in a specific messenger RNA (mRNA) target by Watson-Crick base pairing. The formed DNA-RNA or RNA-RNA duplexes are recognized by cellular enzymes (e.g. RNase-H) that degrade the mRNAs, ultimately leading to a decrease in the cellular levels of the protein.

So far most examples in the literature concerning the use of antisense oligonucleotides in vivo are related to their use in the central nervous system. This may be for several reasons: The CNS is a closed system and as a consequence the injected oligonucleotides remain localized; second, nerve cells may effectively take up oligonucleotides without a delivery system. The dopamine-2 receptor,[49,50] 5-HT6 receptor,[51] the neuropeptide Y-Y1 receptor,[52] and the V1 vasopressin receptor[53] are among the proteins in the CNS that are successfully targeted in an in vivo antisense oligonucleotide approach.

Intracerebroventricular injections of rats with antisense oligonucleotides (AO) directed towards the 5-HT6 receptor caused a specific behavioral syndrome of yawning, stretching and chewing.[51] The absence of specific 5-HT6 ligands made the measurement of the number of 5-HT6 binding sites in these animals difficult. In the frontal lobe the number of 3H-LSD binding sites, as measured in the presence of spiperone for masking binding of the radioligand to 5-HT2a and D2 receptors, was reduced by approximately 20% in the AO treated animals as compared to the scrambled oligonucleotide controls. Since 3H -LSD may label other receptors as well (e.g. the 5-ht5 receptors), the actual decrease in the number of 5-HT6 receptors may be larger following the antisense treatment. Unfortunately 5-HT6 mRNA levels were not measured in these animals. The AO-induced behavioral syndrome could be blocked by atropine, a nonselective muscarinic antagonist. This may suggest that 5-HT6 receptors in brain are involved in the control of cholinergic neurotransmission.

While antisense strategies are conceptually appealing, the interpretation of these experiments is often hampered by the high-degree of non-specificity in these experiments. Oligonucleotides show sequence-specific and non-specific binding to many small molecules and proteins. It may be difficult to separate true antisense effects from the non-specific biological effects of the antisense oligonucleotides (See ref. 54). Also note that the approach is aimed at lowering the target mRNA levels but, in practice, the decrease in specific mRNA levels may be only partial or absent (for instance see ref. 53).

6.6.2 Gene Knockout

The gene knockout strategy enables the generation of mutant mice lacking a specific gene and, therefore, the corresponding protein.[55] This technology relies on the fact that blastocyst derived embryonic stem (ES) cells can be grown in vitro and modified by transfection. In particular, it is possible to disrupt a given gene by homologous recombination. These engineered ES cells are injected into recipient blastocysts and contribute to the formation of a chimeric mouse, and some will transmit the mutation to their germ line. The resulting heterozygous mice are bred to generate mice that are homozygous for the desired mutation.

Mutant mice have been generated that lack the 5-HT1B[56] and the 5-HT2C[57] receptors. These mice do not exhibit any *obvious* developmental defects and are fertile. It is therefore possible to study in these animals all the physiological and behavioral consequences of the absence of the 5-HT1B or 5-HT2C receptors.

The 5-HT1B receptor was the first protein involved in serotonergic signaling to be disrupted.[56] Targeted disruption was achieved by deleting a fragment of the coding sequence; functional inactivation of the 5-HT1B receptor was confirmed by autoradiography with

^{125}I-cyanopindolol. Heterozygote animals expressed the same levels of receptor as wild-types, suggesting that a feedback mechanism may regulate the transcriptional activity of the gene.

Two of the behaviors postulated to be modulated by 5-HT1B receptors were analyzed—locomotion and aggression. Untreated wild-type (WT) and homozygous mutant (5-HT1B -/-) mice were found to display similar levels of locomotor activity in an open field. However, when treated with the 5-HT1 agonist RU24969, the 5-HT1B -/- (homozygous deleted) mice displayed no change in activity whereas wild-type mice doubled their activity. This result confirmed previous pharmacological studies indicating that the locomotor effects of RU24969 are mediated by 5-HT1B receptors.

The aggressiveness of 5-HT1B -/- male mice was assessed by isolating them for four weeks and then exposing them to a non-isolated male WT intruder mouse. The latency and number of attacks displayed by the mice during a 3 minute period were used as indices of aggression. The 5-HT1B -/- mice, when compared with WT mice, exhibited faster, more intense and more frequent attacks. For example, 46% of the 5-HT1B -/- mice attacked the intruder within 10 seconds of introduction, whereas no heterozygous or WT mice attacked during this interval. Tail rattling, an aggressive display, was also more frequent in 5-HT1B -/- mice.

Eltoprazine and fluprazine, drugs classified as serenics because of their antiaggressive properties, are believed to operate as agonists of 5-HT1 receptors. The lack of subtype-specific antagonists has impeded specific identification of the molecular target of serenics. The results with the 5-HT1B -/- mice suggest that the antiaggressive effects of the serenics are mediated, at least in part, by 5-HT1B receptors.

The 5-HT2C knockout was generated by placing a stop codon that eliminated the carboxy terminal portion of the protein.[57] The 5-HT2C receptor is X-linked; brains of mutant males were devoid of 5-HT2C immunoreactivity, and mRNA injected into *Xenopus* oocytes failed to produce functional receptor.

The patterns of expression of 5-HT2C receptors in the hippocampus and spinal cord suggest that they might modulate memory and nociception. However, hippocampal LTP and nociceptive thresholds were the same between wild-type and 5-HT2C -/- mice. Video monitoring revealed that 5-HT2C -/- mice display spontaneous epileptic seizures that were sometimes fatal; survival plots indicated that almost half of the mice had died by 25 weeks of age. Metrazol, a $GABA_A$ receptor antagonist, was used to quantitate seizure susceptibility. Relative to WT mice, the 5-HT2C -/- mice displayed reduced seizure threshold and a more rapid progression through the tonic-clonic phase of the seizure. The unexpected susceptibility of 5-HT2C -/- mice to epileptic seizures raises the possibility that this receptor participates in the modulation of neuronal network excitability.

5-HT2C -/- mice were also noted to be significantly heavier than matched WT controls. Close analysis of their feeding behavior indicated that the weight increase was due entirely to increased intake, and not due to a metabolic derangement that increases the efficiency of caloric storage. The serotonergic agonist 1-(3-chlorophenyl)piperazine (mCPP) drastically reduced food intake by WT mice, but had no effect on food intake of 5-HT2C -/- mice, demonstrating that some appetite suppressants may operate via 5-HT2C receptors.

We must introduce a cautionary note to the studies above since mutant mice develop without the specific protein that has been targeted. As a result, compensatory changes might occur, which will influence the phenotype of the adult animal. An "inducible knockout" strategy that should circumvent the issue of compensatory changes is probably the way to go in the future.

Two different inducible knockout approaches are in development. One makes use of an inducible promotor to control the expression of the Cre recombinase. After induction the Cre recombinase deletes a segment of the targeted gene that is flanked by so-called loxP recombinase recognition sites.[58] The other approach takes advantage of a tetracycline-controlled promoter[59] that was recently shown to yield an inducible inhibition of expression of β-galactosidase in transgenic mice.[60] Variants of the inducible knockout strategies might allow not only inducible, but also tissue specific knockouts.

6.7 CONCLUDING REMARKS

In addition to the receptors described above, a number of 5-HT receptors exist whose pharmacological profile does not correspond to that of any of the cloned 5-HT receptors (called 5-HT1P and 5-HT1S receptors; for a review see ref. 3). It is therefore likely that the number of 5-HT receptors is going to keep increasing.

The availability of the genes encoding all these receptors will aid the design of pharmacological tools, such as selective ligands (using computer aided molecular design techniques), and genetic tools such as gene knockout. A combination of these technologies will hopefully unravel the functions of 5-HT receptors and their role in various physiological pathological states and enhance the prospects for identifying new therapeutic agents to address major unmet medical needs.

REFERENCES

1. Wilkinson LO, Dourish CT. Serotonin and animal behavior. In: Serotonin Receptor Subtypes: Basic and Clinical Aspects. S. Peroutka, ed. New York: John Wiley & Sons Ltd., 1991; 147-210.
2. Sleight AJ, Pierce PA, Schmidt AW, Hekmatpanah CR, Peroutka SJ. The clinical utility of serotonin receptor active agents in neuropsychiatric disease. In: Serotonin Receptor Subtypes: Basic and Clinical Aspects. S. Peroutka, ed. New York: John Wiley & Sons Ltd., 1991; 211-227.

3. Hoyer D, Clarke DE, Fozard JR, Hartig PR, Martin GR, Mylecharane EJ, Sazena PR, Humphrey PPA. VII International union of pharmacology classification of receptors for 5-hydroxytrypamine (serotonin). Pharmacological Reviews 1994; 46:157-203.
4. Martin GR, Humphrey PPA. Receptors for 5-hydroxytryptamine: Current perspectives on classification and nomenclature. Neuropharmacol 1994; 33:261-273.
5. Lucas JJ, Hen R. New players in the 5-ht receptor field: genes and knockouts. Trends Pharmacol Sci 1995; 16:246-252.
6. Peroutka SJ, Howell TA. The molecular evolution of G-protein-coupled receptors: focus on 5-hydroxytryptamine receptors. Neuropharmacol 33:319-324.
7. Gerald C, AdhaM N, Kao H-T, Olsen MA, Laz TM, Schechter LE, Bard JA, Vaysse PJJ, Hartig PR, Branchek TA, Weinshank RL. The 5-ht4 receptor: molecular cloning and pharmacological characterization of two splice variants. EMBO J. 1995; 14:2806-2815.
8. Bockaert J, Fozard JR, Dumuis A, Clarke DE. The 5-ht_4 receptor: a place in the sun. Trends Pharmacol Sci 1992; 13:141-145.
9. Ford AP, Clarke DE. The 5-ht4 receptor. Medicinal Research Reviews 1993; 13:633-662.
10. Grossman CJ, Kilpatrick GJ, Bunce KT. Development of a radioligand binding assay for 5-ht_4 - Receptors in guinea-pig and rat brain. Br J Pharmacol 1993; 109:618-624.
11. Domeneck T, Beleta J, Fernandez AG, Gristwood RW, Cruz Sanchez F, Tolosa E, Palacios JM. Identification and characterization of serotonin 5-ht4 receptor binding sites in human brain: comparison with other mammalian species. Mol Brain Res 1994; 21:176-180.
12. Waeber C, Sebben M, Grossmann C, Javoy-Agra F, Bockaert J, Dumuis A. 3H-GR113808 labels 5-ht4 receptors in the human and guinea pig brain. Neuroreport 1993; 4:1239-1242.
13. Brown AM, Young TJ, Patch TL, Cheung CW, Kaumann AJ, Gaster L, King FD. 125I-SB 207710, A potent, selective radioligand for 5-ht4 receptors. Br J Pharmacol 1993; 110:10P.
14. Reisine, T., Kong, H., Raynor, K., Yano, H., Takeda, J., Yasuda, K., and Bell, G.I. Splice variant of the somatostatin receptor 2 subtype, somatostatin receptor 2B, couples to adenylyl cyclase. Mol Pharmacol 1993; 44(5):1016-1020.
15. Namba T, Sugimoto Y, Negishi M, Irie A, Ushikubi F, Kakizuka A, Ito S, Ichikawa A, Narumiya S. Alternative splicing of C-terminal tail of prostaglandin E receptor subtype EP2 determines G-protein specificity. Nature 1993; 365:166-170.
16. Spengler D, Waeber C, Pantaloni C, Holsboer F, Bockaert J, Seeburg PH, Journot L. Differential signal transduction by five splice variants of the PACAP receptor. Nature 1993; 365:170-175.
17. Plassat JL, Boschert U, Amlaiky N, Hen R. The mouse 5HT_5 receptor reveals a remarkable heterogeneity within the 5HT1D receptor family.

EMBO J 1992; 11:4779-4786.
18. Erlander MG, Lovenberg TW, Baron BM, Delecea L, Danielson PE, Racke M, Slone AL, Siegel BW, Foye PE, Cannon K, Burns JE, Sutcliffe JG. 2 members of a distinct subfamily of 5-Hydroxytryptamine receptors differentially expressed in rat brain. Proc Natl Acad Sci USA 1993; 90:3452-3456.
19. Rees S, den Daas I, Foord S, Goodson S, Bull D, Kilpatrick G, Lee M. Cloning and characterization of the human 5-ht5A serotonin receptor. FEBS Lett 1993; 333:242-246.
20. Matthes H, Boschert U, Amlaiky N, Grailhe R, Plassat JL, Muscatelli F, Mattei MG, Hen R. Mouse 5-Hydroxytryptamine5A and 5-Hydroxytryptamine5B receptors define a new family of serotonin receptors - Cloning, functional expression, and chromosomal localization. Mol Pharmacol 1993; 43:313-319.
21. Wisden W, Parker EM, Mahle CD, Grisel DA, Nowak HP, Yocca FD, Felder CC, Seeburg PH, Voigt MM. Cloning and characterization of the rat 5-ht5B receptor. FEBS Lett. 1993; 333:25-31.
22. Hatziioannou AG, Krauss CM, Lewis MB, Halazonetis TD. Familial holoprosencephaly associated with a translocation breakpoint at chromosomal position 7q36. Am J Med Genet 1991; 40(2):201-205.
23. Pierce PA, Peroutka SJ. Antagonist properties of d-LSD at 5-hydroxytryptamine2 receptors. Neuropharmacology 1990; 3:503-508.
24. Monsma FJ, Shen Y, Ward RP, Hamblin MW, Sibley DR. Cloning and expression of a novel serotonin receptor with high affinity for tricyclic psychotropic drugs. Mol Pharmacol 1993; 43:320-327.
25. Ruat M, Traiffort E, Arrang JM, Tardivellacombe J, Diaz J, Leurs R, Schwartz JC. A novel rat serotonin (5-ht6) receptor - molecular cloning, localization and stimulation of cAMP accumulation. Biochem Biophys Res Commun 1993; 193:268-276.
26. Schoeffter P, Waeber C. 5-Hydroxytryptamine receptors with a 5-ht6 receptor-like profile stimulation adenylyl cyclase activity in pig caudate membranes. Nauyn-Schmiedeberg's Arch Pharmacol 1994; 350:356-360.
27. Conner DA, Mansour TE. Serotonin receptor-mediated activation of adenylate cyclase in the neuroblastoma NCB.20: a novel 5-hydroxytryptamine receptor. Mol Pharmacol 1990; 37:742-751.
28. Unsworth CD, Molinoff PB. Characterization of the 5-Hydroxytryptamine receptor in mouse neuroblastoma N18TG2 cells. J Pharm Exp Ther 1994; 269:246-255.
29. Roth BL, Craigo SC, Choudhary M, Uluer A, Monsma FJ, Shen Y, Meltzer HY, Sibley DR. Binding of typical and atypical antipsychotic agents to 5-Hyroxytryptamine-6 and 5-hydroxytryptamine-7 receptors. J Pharm Exp Ther 1994; 268:1403-1410.
30. Bard J, Zhombick J, Adham N, Vaysse P, Branchek T, Weinshank R. Cloning of a novel human serotonin receptor (5-ht7) positively linked to adenylate. J Biol Chem 1993; 268:23422-23426.

31. Lovenberg TW, Baron BM, de Lecea L, Miller, JD, Prosser RA, Rea MA, Foye PE, Racke M, Slone AL, Siegel BW et al. A novel adenylyl cyclase-activating serotonin receptor (5-ht7) implicated in the regulation of mammalian circadian rhythms. Neuron 1993; 11:449-458.
32. Meyerhof W, Obermuller F, Fehr S, Richter D. A novel rat serotonin receptor - primary structure, pharmacology, and expression pattern in distinct brain regions. DNA Cell Biol 1993; 12:401-409.
33. Ruat M, Traiffort E, Leurs R, Tardivel-Lacombe J, Diaz J, Arrang JM, Schwartz JC. Molecular cloning, characterization and localization of a high affinity serotonin receptor (5-ht7) activating cAMP formation. Proc Natl Acad Sci USA 1993; 90:8547-8551.
34. Shen Y, Monsma FJ, Metcalf MA, Jose PA, Hamblin MW, Sibley DR. Molecular cloning and expression of a 5-Hydroxytryptamine7 serotonin receptor subtype. J Biol Chem 1993; 268:18200-18204.
35. Plassat JL, Amlaiky N, Hen R. Molecular cloning of a mammalian serotonin receptor that activates adenylate cyclase. Mol Pharmacol 1993; 44:229-236.
36. Tsou AP, Kosaka A, Bach C, Zuppan P, Yee C, Tom L, Alvarez R, Ramsey S, Bonhaus DW, Stefanisch E et al. Cloning and expression of a 5-hydroxytryptamine7 receptor positively coupled to adenylyl cyclase. J Neurochem 1994; 63, 456-464.
37. Nelson CS, Cone RD, Robbins LS, Allen CN, Adelman JP. Cloning and expression of a 5HT7 receptor from Xenopus laevis. Receptors and Channels 1995; 3:61-70.
38. Witz P, Amlaiky N, Plassat JL, Maroteaux L, Borrelli E, Hen R. Cloning and characterization of a Drosophila serotonin receptor that activates adenylate cyclase. Proc Natl Acad Sci USA 1990; 87:8940-8944.
39. Saxena PR, Mylecharane EJ, Heiligers J. Analysis of the heart rate effects of 5-Hydroxytryptamine in the cat: mediation of tachycardia by 5-ht_1-like receptors. Naunyn Schmiedeberg's Arch Pharmacol 1985; 330:121-129.
40. Connor HE, Feniuk W, Humphrey PPA, Perren MJ. 5-Carboxamidotryptamine is a selective agonist at 5-Hydroxytryptamine receptors mediating vasodilatation and tachycardia in anaesthetized cats. Br J Pharmacol 1986; 87:417-426.
41. Trevethick MA, Feniuk W, Humphrey PPA. 5-Carboxamidotryptamine: a potent agonist mediating relaxation and elevation of cyclic AMP in the isolated neonatal porcine vena cava. Life Sci 1986; 38:1521-1528.
42. Sleight AJ, Carolo C, Petit N, Zwingelstein C, Bourson A. Identification of 5-hydroxytryptamine7 receptor binding sites in rat hypothalamus: sensitivity to chronic antidepressant treatment. Mol Pharmacol 1995; 47:99-103.
43. Shenker A, Maayani S, Weinstein H, Green JP. Two 5-ht receptors linked to adenylate cyclase in guinea pig hippocampus are discriminated by 5-carboxamidotryptamine and spiperone. Eur J Pharmacol 1985; 109:427-429.

44. Markstein R, Hoyer D, Engel G. 5-ht1A receptors mediate stimulation of adenylate cyclase in rat hippocampus. Naunyn Schmiedeberg's Arch Pharmacol 1986; 333:335-345.
45. Fayolle C, Fillion MP, Barone P, Oudar P, Rousselle JC, Fillion G. 5-Hydroxytryptamine stimulates two distinct adenylate cyclase activities in rat brain: high affinity activation is related to a 5-ht1 subtype different from 5-ht1A, 5-ht1B, 5-ht1C. Fund Clin Pharmacol 1988; 2:195-214.
46. Prosser RA, Dean RR, Edgar DM, Heller HC, Miller JD. Serotonin and the mammalian circadian system. I. *In vitro* phase shifts by serotonergic agonists and antagonists. J Biol Rhythms 1993; 8:1-16.
47. Lucas JJ, Mellstrom B, Colado MI, Naranjo JR. Molecular Mechanisms of Pain - Serotonin(1A) Receptor Agonists Trigger Transactivation by c-fos of the Prodynorphin Gene in Spinal Cord Neurons. Neuron 1993; 10:599-611.
48. Kawahara F, Saito H, Katsuki H. Inhibition by 5-ht7 receptor stimulation of GABAA receptor-activated current in cultured rat suprachiasmatic neurones. J Physiol 1994; 478:67-73.
49. Zhou LW, Zhang SP, Qin ZH, Weiss B. *In vivo* administration of an oligodeoxynucleotide antisense to the D2 dopamine receptor messenger RNA inhibits D2 dopamine receptor-mediated behavior and the expression of D2 dopamine receptors in mouse striatum. J Pharmacol Exp Ther 1994; 268, 1015-1023.
50. Silvia CP, King GR, Lee TH, Xue ZY, Caron MG, Ellinwood EH. Intranigral administration of D2 dopamine receptor antisense oligodeoxynucleotides establishes a role for nigrostriatal D2 autoreceptors in the motor actions of cocaine. Mol Pharmacol 1994; 46:51-57.
51. Bourson A, Borroni E, Austin RH, Monsma FJ, Sleight AJ. Determination of the role of the 5-ht6 receptor in the rat brain: A study using antisense oligonucleotides. J Pharm Exp Ther 1995; 274:173-180.
52. Wahlestedt C, Merlo Pich E, Koob GF, Yee F, Heilig M. Modulation of anxiety and neuropeptide Y-Y1 receptors by antisense oligodeoxynucleotides. Science 1993; 259:528-531.
53. Landgraf R, Gerstberger R, Montkowski A, Probst JC, Wotjak CT, Holsboer F, Engelmann M. V1 vasopressin receptor antisense oligodeoxynucleotide into septum reduces vasopressin binding, social discrimination abilities, and anxiety-related behavior in rats. J Neurosci 1995; 15:4250-4258.
54. Wagner R. Gene inhibition using antisense oligodeoxynucleotides. Nature 1994; 372:333-335.
55. Thomas KR, Capecchi MR. Targeted disruption of the murine *int-1* protooncogene resulting in severe abnormalities in midbrain and cerebellar development. Nature 1993; 346:847-850.
56. Saudou F, Ait Amara D, LeMeur M, Ramboz S, Segu L, Buhot MC, Hen R. Enhanced aggressive behavior in mice lacking 5-ht1B receptor. Science 1994; 265:1875-1878.

57. Tecott LH, Sun LM, Akana SF, Strack AM, Lowenstein DM, Dallmann MF, Julius D. Eating disorder and epilepsy in mice lacking 5-ht2C serotonin receptors. Nature 1995; 374:542-546.
58. Kühn R, Schwenk F, Aguet M, Rajewsky K. Inducible gene targeting in mice. Science 1995; 269:1427-1429.
59. Gossen M, Bujard H. Tight control of gene expression in mammalian cells by tetracycline-responsive promoters. Proc Natl Acad Sci USA 1992; 89:5547-5551.
60. Furth PA, St. Onge L, Boger H, Gruss P, Gossen M, Kistner A, Hennighausen L. Temporal control of gene expression in transgenic mice by a tetracycline responsive promoter. Proc Natl Acad Sci USA 1994; 91:9302-9306.

CHAPTER 7

Recombinant 5-Hydroxytryptamine3 Receptors

John A. Peters, Anthony G. Hope, Lesley Sutherland and Jeremy J. Lambert

7.1 INTRODUCTION

The discovery of a 5-HT receptor subtype ('M' receptor) that some thirty years later would emerge as the 5-HT3 receptor, occurred in the early 1950s as a result of classical pharmacological studies conducted upon the guinea-pig ileum.[1,2] Unfortunately, for the next two decades, characterization of the receptor and an assessment of its potential role in physiological and pathophysiological processes was severely impeded by a lack of selective antagonist and agonist compounds. Despite this handicap, detailed studies of the actions of 5-HT in a variety of peripheral tissue preparations gradually provided evidence that neuronally located excitatory 5-HT receptors akin, or identical, to the 'M' receptor were widely distributed in the autonomic and sensory divisons of the peripheral nervous system.[3,4] Consistent with the original report of Gaddum and Picarelli,[2] cocaine was frequently described as an effective, albeit poorly selective, antagonist. Both in terms of definitive characterization of the receptor and subsequent drug development, a key event was the demonstration by Fozard and Mobarok Ali that the anti-emetic dopamine D2 antagonist, metoclopramide, additionally acted as an effective and surmountable antagonist of the neuronal 5-HT receptor.[5] Indeed, the marriage of structural elements present in metoclopramide and cocaine led, *via* a series of substituted benzoic acid esters of tropane, to the synthesis of bemesetron (MDL-72222), the first selective antagonist of the receptor.[6-8] Contemporaneously, researchers at Sandoz followed the well established strategy of

Recombinant Cell Surface Receptors: Focal Point for Therapeutic Intervention, edited by Michael J. Browne.

chemical substitution and derivatization of the agonist itself, eventually synthesizing both a relatively selective 5-HT3 receptor agonist (2-methyl-5-HT) and a series of highly selective antagonists that included tropisetron (ICS 205-930).[9]

The advent of the aforementioned compounds allowed Bradley et al[10] to apply rigid criteria to the definitition of 5-HT3 receptor mediated responses. These included susceptibility to antagonism by bemesetron and tropisetron, resistance to blockade by compounds acting as antagonists at a variety of G-protein-coupled 5-HT receptors (e.g. methysergide, methiothepin, ketanserin) and stimulation by 2-methyl-5-HT. An additional criterion, albeit a negative one, was lack of agonism by 5-methoxytryptamine.[10] Of equal importance, as the number of selective 5-HT3 receptor antagonists steadily grew to include compounds such as ondansetron (GR 38032)[11] and granisetron (BRL 43694),[12] it became possible to investigate the role of the 5-HT3 receptor in the functioning of the nervous system.

Interest in the 5-HT3 receptor field intensified when 5-HT3 recognition sites were localized in the brain of both rodents and humans by the use of radiolabelled 5-HT3 receptor antagonists.[13-16] That demonstration provided anatomical support for an emerging behavioral literature suggesting 5-HT3 receptor antagonists possess wide-ranging psychoactive properties.[17] Thus, in animal models, 5-HT3 receptor antagonists exerted effects predictive of anxiolytic,[18] antipsychotic,[19] and cognitive enhancing[20] activities. Additionally, this class of compound has been suggested to possess potential in the treatment of drug dependence and abuse.[21] However, despite these tantilizing predictions, the therapeutic use of 5-HT3 receptor antagonists is currently limited to the suppression of acute nausea and vomiting occurring as a consequence of anticancer chemo- and radiotherapy, or in response to surgical interventions involving the use of a general anaesthetic.[22]

7.2 ISOLATION OF CLONED 5-HT3 RECEPTOR SUBUNITS

5-HT3 receptors, in contrast to all other 5-HT receptor subtypes which couple to G-proteins, belong to a family of ligand-gated ion channels that includes the nicotinic, $GABA_A$ and glycine receptors.[23,24] The functional properties of 5-HT3 receptors endogenous to peripheral and central neurones and a variety of neuronal clonal cell lines have been the subject of several recent and extensive reviews[25-29] and will be described only in comparision to data obtained from heterologously expressed recombinant 5-HT3 receptors.

To date, only one gene encoding a 5-HT3 receptor subunit has been identified, although species variants and alternatively spliced isoforms of the subunit have been isolated (Table 7.1). This is in marked contrast to other ligand-gated ion channels, which exist as large families of receptor subunits encoded by separate genes. The initial isolation of a cDNA encoding a 5-HT3 receptor subunit[30] was made possible by exploiting the murine hydridoma cell line NCB-20 which had previously been documented to possess a high density of functional 5-HT3 receptors.[23,31] A size selected cDNA library was constructed and screened for the expression of 5-HT-gated currents in *Xenopus laevis* oocytes. A single 2131 base pair cDNA (m5-HT3R-A) was identified as encoding a functional homo-oligomeric receptor, sharing many of the properties associated with the 5-HT3 receptor native to NCB-20 cells.[30] The 489 amino acid protein (originally reported to be 487 due to errors in the nucleotide sequence; see Fig. 7.1)[32,33] coded by the 5-HT3R-A cDNA corresponds to a polypeptide with a molecular weight of 56,222 Da, although post-translational cleavage of a 23 amino acid

Table 7.1. Sources and structural characteristics of recombinant 5-HT3R-A species homologs and splice-variants.

Classification	Species & DNA Source	Predicted polypeptide	Accession Number	Ref.
m5-HT3R-A_L	Mouse NCB 20 hydridoma cell line expression library	487 aa (23 aa signal peptide) mw calc. 53 509.	M74425	30
m5-HT3R-A_S	Mouse N1E-115 neuroblastoma cell line	483 aa (23 aa signal peptide) mw calc. 53 178.	X72395	32
r5-HT3R-A_S	Rat superior cervical ganglion library	462 aa (incomplete signal peptide of 2 aa) mw calc. 53 062.	U01227	37
r5-HT3R-A_S	Rat brain	483 aa (23 aa signal peptide) mw calc. 53 017.	D49395	39
h5-HT3R-A_S	Human hippocampus	478 aa (23 aa signal peptide) mw calc. 52 789.	D49394	39
h5-HT3R-A_S	Human amygdala	478 aa (23 aa signal peptide) mw calc. 52 778.	Not available	38

Notes: Nucleotide sequences of cDNAs encoding the above subunits are available using the Genbank accession numbers listed. Calculated molecular weights (mw calc.) refer to the predicted mature protein. Abbreviation: aa = amino acid.

```
h5-HT3R-AS   MLLWVQQALL ALLLPTLLAQ GEARRSR--- --NTTRPALL RLSDYLLTNY  22
m5-HT3R-AS   MRLCIPQVLL ALFLSMLTAP GEGSRRRATQ ARDTTQPALL RLSDHLLANY  27
m5-HT3R-AL   MRLCIPQVLL ALFLSMLTAP GEGSRRRATQ ARDTTQPALL RLSDHLLANY  27
r5-HT3R-AS                           KGSRRRATQ AHSTTQPALL RLSDHLLANY  27
                    Signal Sequence                  ▼               ▼

h5-HT3R-AS   TKGVRPVRDW RKPTTVSIDV IVYAILNVDE KNQVLTTYIW YRQYWTDEFL  72
m5-HT3R-AS   KKGVRPVRDW RKPTTVSIDV IMYAVLNVDE KNQVLTTYIW YRQYWTDEFL  77
m5-HT3R-AL   KKGVRPVRDW RKPTTVSIDV IMYAILNVDE KNQVLTTYIW YRQYWTDEFL  77
r5-HT3R-AS   KKGVRPVRDW RKPTLVSIDV IMYAILNVDE KNQVLTTYIW YRQFWTDEFL  77

h5-HT3R-AS   QWNPEDFDNI TKLSIPTDSI WVPDILINEF VDVGKSPNIP YVYIRHQGEV 122
m5-HT3R-AS   QWTPEDFDNV TKLSIPTDSI WVPDILINEF VDVGKSPNIP YVYVHHRGEV 127
m5-HT3R-A    QWTPEDFDNV TKLSIPTDSI WVPDILINEF VDVGKSPNIP YVYVHHRGEV 127
r5-HT3R-AS   QWTPEDFDNV TKLSIPTDSI WVPDILINEF VDVGKSPSIP YVYVHHQGEV 127
                     ▲

h5-HT3R-AS   QNYKPLQVVT ACSLDIYNFP FDVQNCSLTF TSWLHTIQDI NISLWRLPEK 172
m5-HT3R-AS   QNYKPLQLVT ACSLDIYNFP FDVQNCSLTF TSWLHTIQDI NITLWRSPEE 177
m5-HT3R-AL   QNYKPLQLVT ACSLDIYNFP FDVQNCSLTF TSWLHTIQDI NITLWRSPEE 177
r5-HT3R-AS   QNYKPLQLVT ACSLDIYNFP FDVQNCSLTF TSWLHTIQDI NISLWRTPEE 177
                         S            ▲S                      ▲

h5-HT3R-AS   VKSDRSVFMN QGEWELLGVL PYFREFSMES SNYYAEMKFY VVIRRRPLFY 222
m5-HT3R-AS   VRSDKSIFIN QGEWELLEVF PQFKEFSIDI SNSYAEMKFY VIIRRRPLFY 227
m5-HT3R-AL   VRSDKSIFIN QGEWELLEVF PQFKEFSIDI SNSYAEMKFY VIIRRRPLFY 227
r5-HT3R-AS   VRSDKSIFIN QGEWELLGVF TKFQEFSIET SNSYAEMKFY VVIRRRPLFY 227
                                                              _____

h5-HT3R-AS   VVSLLLPSIF LMVMDIVGFY LPPNSGERVS FKITLLLGYS VFLIIVSDTL 272
m5-HT3R-AS   AVSLLLPSIF LMVVDIVGFC LPPDSGERVS FKITLLLGYS VFLIIVSDTL 277
m5-HT3R-AL   AVSLLLPSIF LMVVDIVGFC LPPDSGERVS FKITLLLGYS VFLIIVSDTL 277
r5-HT3R-AS   AVSLLLPSIF LMVVDIVGFC LPPDSGERVS FKITLLLGYS VFLIIVSDTL 277
             ____________________________   _______________________
                         M1                          M2

h5-HT3R-AS   PATAIGTPLI GVYFVVCMAL LVISLTETIF IVRLVHKQDL QQPVPAWLRH 322
m5-HT3R-AS   PATAIGTPLI GVYFVVCMAL LVISLAETIF IVRLVHKQDL QRPVPDWLRH 327
m5-HT3R-AL   PATAIGTPLI GVYFVVCMAL LVISLAETIF IVRLVHKQDL QRPVPDWLRH 327
r5-HT3R-AS   PATAIRTPLI GVYFVVCMAL LVISLAETIF IVQLVHKQDL QRPVPDWLRH 327
                 ____________________
                          M3

h5-HT3R-AS   LVLERIAWLL CLREQSTSQR PPATSQATKT DDCS------ AMGNHCTHMG 366
m5-HT3R-AS   LVLDRIAWIL CLGEQPMAHR PPATFQANKT DDCS------ AMGNHCSHVG 371
m5-HT3R-AL   LVLDRIAWIL CLGEQPMAHR PPATFQANKT DDCSGGSDLLP AMGNHCSHVG 377
r5-HT3R-AS   LVLDRIAWLL CLGEQPMAHR PPATFQANKT DDCS------ AMGNHCSHVG 371
                                                  ======

h5-HT3R-AS   GPQDFEKSPR DRCSPPPPPR EASLAVCGLL QELSSIRQFL EKRDEIREVA 416
m5-HT3R-AS   GPQDLEKTPR GRGSPLPPPR EASLAVRGLL QELSSIRHFL EKRDEMREVA 421
m5-HT3R-AL   GPQDLEKTPR GRGSPLPPPR EASLAVRGLL QELSSIRHFL EKRDEMREVA 425
r5-HT3R-AS   SPQDLEKTSR SRDSPLPPPR EASLAVRGLL QELSSIRHSL EKRDEMREVA 421

h5-HT3R-AS   RDWLRVGSVL DKLLFHIYLL AVLAYSITLV MLWSIWQYA 455
m5-HT3R-AS   RDWLRVGYVL DRLLFRIYLL AVLAYSITLV TLWSIWHYS 460
m5-HT3R-AL   RDWLRVGYVL DRLLFRIYLL AVLAYSITLV TLWSIWHYS 466
r5-HT3R-AS   RDWLRVGYVL DTLLFRIYLL AVLAYSITLV TLWSIWHYS 460
                                  _____________________
                                           M4
```

signal sequence,[30] at the amino terminus of the putative precursor polypeptide predicts the molecular weight of the mature protein subunit to be 53,766 Da. The calculated size of the cloned 5-HT3 receptor subunit is consistent with the identification of the 54 kDa and 50 kDa subunits determined by SDS PAGE analysis of the native 5-HT3 receptor purified from NCB-20 cells[31] and NG108-15 cells[34], respectively.

Subsequently, a 5-HT3 receptor cDNA was isolated from the murine neuroblastoma N1E-115 cell line, using the polymerase chain reaction (PCR).[32] Transient expression of the cRNA in *Xenopus laevis* oocytes,[32] or its stable expression in HEK-293 cells,[35] resulted in the synthesis of functional homo-oligomeric 5-HT3 receptors. The polypeptide predicted by the cDNA isolated by Hope et al[32] demonstrated extensive sequence identity (98%) to that of the previously cloned 5-HT3 receptor subunit,[30] the most significant difference being the deletion of six consecutive amino acids within a putative cytoplasmic loop. It was proposed,[32] and subsequently confirmed,[33] that the expression of the two cDNAs results from alternative RNA splicing of a single gene product that gives rise to a 'long' (5-HT3R-A_L) form of the subunit which contains the hexapeptide and a 'short' (5-HT3R-A_S) form that lacks it (see section 7.3).

In addition to the mouse 5-HT3 receptor subunit, a species homolog of 5-HT3R-A has been cloned from rat superior cervical ganglion cDNA libraries by two groups. Johnson and Heinemann[36] reported the isolation of a full-length cDNA which directed the expression of a functional homo-oligomeric receptor complex in *Xenopus* oocytes. Isenberg et al[37] screened their library with oligonucleotides used to clone nicotinic acetylcholine receptor subunits and obtained a partial cDNA which lacked the signal sequence, but was otherwise complete. The cDNA was predictive of a mature protein of 460 amino acids (Fig. 7.1) exhibiting 95% amino acid sequence identity to the m5-HT3R-A_L.[37] Expression of the latter clone in *Xenopus* oocytes has been achieved by ligating a *Cla I* fragment of the partial cDNA to the 5′ *Cla I* fragment of the m5-HT3R-A (J Yang, A Hope, J Lambert and I Mair, unpublished observations).

Most recently, a cDNA encoding a human homolog of 5HT3R-A has been isolated by two groups. Belelli et al[38] isolated a cDNA from a library constructed using mRNA prepared from human amygdala,

Fig. 7.1. (opposite page) Alignment of the predicted amino acid sequences of the human (h), mouse (m) and rat (r) 5-HT_3R-A_S subunits, together with the splice variant m5-HT_3R-A_L. The putative membrane spanning segments (M1-4) are singly underlined and the additional hexapeptide within m5-HT_3R-A_L doubly underlined. The two cysteine residues which enclose the loop stucture within the extracellular N-terminal domain are indictated (S). Conserved residues representing potential sites of N-linked glycosylation are denoted (▲); note the two additional sites (▼) in h5-HT_3R-A_S. Conserved phosphorylation consensus sequences for A-kinase (●) and casein kinase II (◆) are indicated, together with a sequence for phosphorylation by tyrosine kinase (■), which is absent from the human isoform.

whereas Miyake et al[39] applied a combination of reverse transcriptase (RT)-PCR and rapid amplification of DNA ends (RACE) to construct a 5-HT3 receptor cDNA from human hippocampal mRNA. The cDNAs predict a mature polypeptide comprising 454 amino acids. Once again, sequence homology to other cloned 5-HT3 receptor subunits is high, sharing 85% and 84% amino acid sequence identity with the m5-HT3R-A_S and r5-HT3R-A_S subunits, respectively.

5-HT3R-A subunits expressed heterologously as a homo-oligomeric receptor complex possess biophysical and pharmacological properties that closely resemble those of native 5-HT3 receptors. In common with other members of the Cys-loop ligand-gated ion channel family,[40,41] native 5-HT3 receptors are predicted to be hetero-oligomeric proteins, constructed from a pentameric array of constituent receptor subunits. Electron microscopy of both the recombinant m5-HT3R-A_S subunit expressed in insect sf9 cells,[42] and the 5-HT3 receptor native to NG108-15 cells,[34,43] has provided evidence for this proposed quaternary structure. 5-HT3 receptor protein purified from both systems and negatively stained with uranyl acetate was observed as rosette-shaped structures with a diameter of 7-9 nm enclosing a centrally stained pit of 2-3 nm, which presumably represents the pore-forming ion-channel. A side view of these structures reveals cylindrical particles 11-14 nm in length. Whether hetero-pentameric 5-HT3 receptors exist in nature is a matter for speculation. Cloning studies have, to date, failed to identify novel 5-HT3 receptor cDNAs. Additionally, low stringency Southern hybridization to murine genomic DNA, using 5HT3R-A cDNA sequences as a probe, did not detect the presence of any closely related subunits (AG Hope, unpublished observations) although the presence of more distantly related sequences can not, of course, be excluded by this analysis. In this respect, it is interesting that purified 5-HT3 receptor protein resolved by SDS PAGE reveals a 38 kDa band in addition to the aforementioned 50-54 kDa subunit.[32] Additionally, an antibody raised to the large cytoplasmic loop (see below) of the 5-HT3R-A recognizes a protein in rat brain and NG 108-15 cells with a molecular mass of 35 kDa determined by western blotting with anti-5-HT3 receptor anti-bodies.[44] Provided that such results do not merely reflect degradation of 5-HT3 receptor protein, they may point to an as yet undetected subunit.

A comparison of the primary structure of the 5-HT3R-A subunit with that determined for other ligand-gated ion-channels assigns it to a family that includes the nAChR, $GABA_A$ and glycine receptors. The extracellular N-terminal domain associated with each of these receptors has the common structural feature of two cysteine residues separated by 13 amino acids.[41] Within this Cys-loop family, the m5-HT3R-A subunit is most closely related to the cation-selective nicotinic subunits, sharing 27% sequence identity with the α subunit of *Torpedo* and 30% identity with the $\alpha 7$ subunit of the chick neuronal nACh receptor.[30] Very little is known of the tertiary structure of the 5-HT3 receptor and as a consequence the model proposed for the nAChR is currently employed.[30] The hydrophobicity profile[45] of the 5-HT3R-A subunit indicates the presence of four hydrophobic domains (M1-M4). Although the latter fulfil the requirements for the formation of membrane spanning α helices,[45,46] there is as yet no direct evidence to verify this secondary structure. In this respect, it is noteworthy that recent data obtained for the nAChR of *Torpedo* by a combination of electron microscopy[47] and Fourier transformed infra red spectroscopy[48] are consistent with the contribution of both α-helical and β-strand and turn structures to the intramembranous domains, with α-helical structures almost certainly contibuting to M2[41,48] and probably M4.[48] By analogy to the other members of the Cys-loop receptor family, the M2 domain is thought to contribute to the formation of the cation conducting ion channel.[41,49] Features within, and adjacent to, the M2 domain that are conserved between nAChR and 5-HT3R-A subunits include: (i) the presence of acidic amino acid residues which form outer, intermediate and inner rings of negative charge within the pentameric complex; these bracket the M2 sequence; (ii) a characteristic repetition of polar amino acids that are postulated to face the channel lumen to give rise to serine and threonine rings, and (iii) an equatorial ring of leucine residues located at the approximate midpoint of M2[49] (Figs. 7.1 and 7.2). Site-directed mutagenesis of nicotinic receptor complexes has revealed the importance of these conserved features as determinants of ion permeation and conduction, desensitization and channel gating.[41,49] However, data relating to the role of these residues in 5-HT3 receptor function is presently limited to the equatorial leucine ring[50] (see section 7.7). Peculiarly, the 5-HT3R-A amino acid sequence within the M2 domain indicates the presence of a basic lysine residue which is predicted to reside at a distance approximately one-third into the membrane from the cytoplasmic surface. This feature is unique to the 5-HT3R-A receptor channel complex (see section 7.7).

At the amino terminus of 5-HT3R-A, a large extracellular domain exists which contains, or contributes to, the ligand binding domain.[51] The latter contains a number of putative sites for N-linked glycosylation,[52] the absolute number and position of which vary from one species to another.[38] This is consistent with the glycoprotein nature of both purified 5-HT3 receptors[31] and the m5-HT3R-A.[42] A second hydrophilic domain links the M3 and M4 domains. This region, which is predicted to exhibit an intracellular location, contains consensus phosphorylation sites for casein kinase II, tyrosine kinase and protein kinase A which may be involved in the regulation of 5-HT3 receptor function and expression[30] (Fig. 7.1).

7.3 ALTERNATIVE SPLICING

Apart from the species dependent isoforms of 5-HT3R-A described above, alternative splicing of the primary 5-HT3R-A transcript represents the only established means of generating structural diversity within

	-5'	-4'	-3'	-2'	-1'	0	1'	2'	3'	4'	5'	6'	7'	8'	9'	10'	11'	12'	13'	14'	15'	16'	17'	18'	19'	20'	21'	22'	23'
	Cytoplasmic			Intermediate																					Extracellular				
	=				=					▼																=			
m5-HT$_3$R-A	D	S	-	G	E	R	V	S	F	K	I	T	L	L	L	G	Y	S	V	F	L	I	I	V	S	D	T	L	P
Chick nAChR α7	D	S	-	G	E	K	I	S	L	G	I	T	V	L	L	S	L	T	V	F	M	L	L	V	A	E	I	M	P
Mouse nAChR α1	D	S	-	G	E	K	M	T	L	S	I	S	V	L	L	S	L	T	V	F	L	L	V	I	V	E	L	I	P
					▲			▲							◆				◆										

Fig. 7.2. Alignment of the amino acid sequence of the M2 segment and immediately adjacent regions of m5-HT$_3$R-A$_S$, with the corresponding regions of cloned nAChR subunits. Conserved negatively charged amino acids (Imoto et al[137]) are indicated by (=). Note that the cytoplasmic ring of charge is absent from h5-HT$_3$R-A$_S$ due to the substitution of D by N.[38,39] Amino acid residues which are suggested to constitute the selectivity filter of nicotinic ACh-gated ion channels are indicated by (▲).[138,139] A conserved leucine residue, which in the chick α7 homo-oligomeric nAChR) is implicated in determining both desensitization kinetics and pharmacological properties[127,140] is denoted by (◆). Also indicated as (◆) is a conserved valine residue that influences agonist-induced densensitization in the nAChR α7 subunit.[141] A positively charged lysine residue that is unique to the 5-HT$_3$R-A subunit is indicated as (▼).

this class of receptor. As mentioned above, a splice variant of 5-HT3R-A was first identified by Hope et al[32] and was shown to assemble efficiently to form functional homo-oligomeric 5-HT3 receptors in both *Xenopus laevis* oocytes and HEK-293 cells.[35] The most striking difference between the two cloned 5-HT3 receptor subunits is the deletion of six amino acids (GSDLLP) from the putative large intracellular loop of the cloned subunit (i.e. m5-HT3R-A_S) isolated from N1E-115 cells[32] (Fig. 7.1). Analysis of total RNA isolated from both N1E-115 cells and NCB-20 cells revealed that the two cloned 5-HT3 receptor subunits are co-expressed in both cell lines.[32] The two 5-HT3 receptor subunits were subsequently also identified in the murine hybridoma NG108-15 cell line.[33] Using PCR in combination with nucleotide sequence analysis to characterize a population in excess of one hundred 5-HT3 receptor cDNAs isolated from a NG108-15 cDNA library, 80% of the clones were found to correspond to the m5-HT3R-A_S subunit.[33] This observation was further supported by RNase protection assays, which also confirmed m5-HT3R-A_S as being the predominant mRNA species encoding the 5-HT3 receptor, the short form of the clone being expressed in 4-6 fold excess when compared to m5-HT3R-A_L transcripts in neuroblastoma cells, as well as in neurones of the hippocampus, cortex and superior cervical ganglion of the mouse.[33] It has been noted that a putative casein kinase II site occurs in 5-HT3R-A_L at the proximal splice junction.[33] This consensus sequence is disrupted in 5-HT3R-A_S.

In common with the m5-HT3R-A subunit, splice variants of the rat cloned 5-HT3 receptor subunit also exist. In this case, the 'long' and 'short' forms of the clone are reported to differ by the incorporation or deletion of 5 amino acid residues (GSLLP), respectively, in the second large intracellular loop.[53] Neither clone presents the casein kinase II consensus sequence. However, a six amino acid deletion identical to that occurring in m5-HT3R-A_S has been quoted by others.[54] Both splice variants have been identified as existing in central and peripheral neurones of the rat, as determined using PCR.[53] In addition, while the short form appears to be the predominant variant, the two splice variants appear to be differentially regulated during development[53,55] (see section 7.5).

7.4. ORGANIZATION AND EVOLUTION OF THE 5-HT3R-A GENE

Two independent reports have described the cloning of genomic DNA fragments which span the coding exons of the 5HT3R-A gene.[33,54] Overlapping fragments were cloned from phage genomic DNA libraries using labelled cDNA sequences as probes. The source of the genomic DNA was BALB/C mice[54] and D3 ES cells.[33]

The 5HT3R-A gene is divided into 9 coding exons which span a total of 12 Kb of genomic sequence (Fig. 7.3). The positions of the intron/exon boundaries reported by the two groups[33,54] are in agreement.

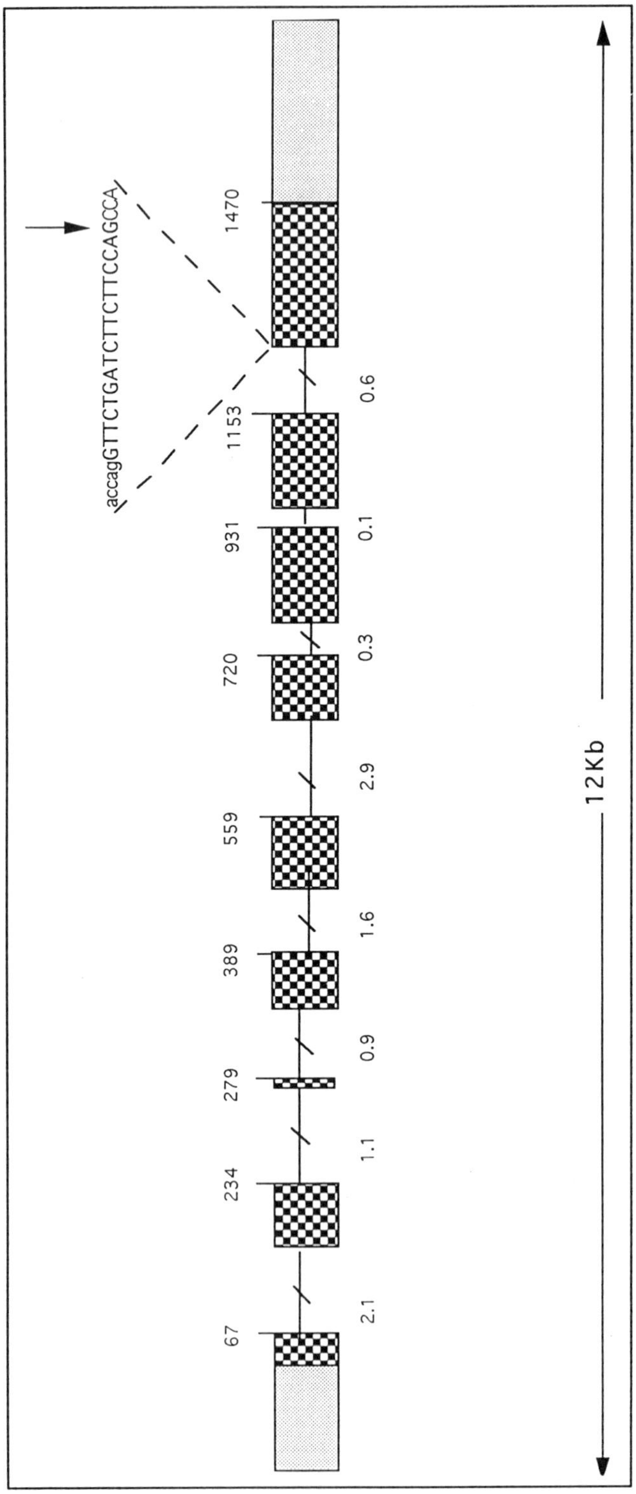

Fig. 7.3. Structure of the $5HT_3R\text{-}A_S$ gene. Exonic sequences are shown as boxes and intronic sequences as lines. The nucleotide coordinates of the splice junctions (assigning the "A" of the initiation codon as nucleotide 1) are indicated, as are the sizes (in kb) of the intervening sequences. Protein coding sequences of the exons are represented as checked areas while non-coding sequences are stippled. Also represented is the alternative splice site within exon nine which is used to generate the two forms of the $5HT_3R\text{-}A$ receptor. Intronic sequence is represented as lower case and exonic as upper case with the position of the alternative splice site indicated by the arrow.

The genomic DNA sequence would seem to confirm that the observed variation within 5HT3R-A is the result of alternative splicing, as exon nine contains two possible splice acceptor sites, 18 bp apart, which could generate the two forms of the subunit. The position of the two possible acceptor splice sites within exon nine are indicated in Figure 7.3. Studies showing that the mRNA encoding 5HT3R-A_S is the predominant species (see sections 7.3 and 7.5) suggests that the downstream acceptor site is favored, although the mechanism governing the choice of splice site is unknown. In a recent study,[55] where 5-HT3R-A mRNA present in NG 108-15 cells was studied by quantitative PCR, the ratio of 'short' to 'long' (S/L) forms was found to be influenced by external factors that influence the differentiation of this cell line. In brief, agents increasing the intracellular concentration of cyclic AMP depressed 5-HT3R-A mRNA levels and increased the S/L ratio. Conversely, serum deprivation, in conjunction with growth factors (nerve growth factor (NGF); basic fibroblast growth factor or epidermal growth factor), decreased both 5-HT3R-A mRNA expression and the S/L ratio.[55] It should be noted that in PC12 cells, NGF in the presence of serum elevates 5-HT3R-A mRNA levels,[37] a result corroborated by increased expression of 5-HT3 receptors detected both in radioligand binding[56] and electrophysiological[37,57] studies.

PCR amplification of a region of human genomic DNA which spans the exon nine sequence has shown that the equivalent exon in human DNA does not contain the proximal splice acceptor site, indicating that only the short form of 5HT3R-A will be found in this species.[33] Although the functional significance of the two possible 5HT3R-A subunits is debatable (see section 7.6) the data from the human genomic sequence would suggest that the long form is not essential, at least not in all species.

Uetz et al[54] have used Southern hybridization to genomic DNA isolated from a panel of rodent/human somatic cell hybrids to assign the human 5HT3R-A gene to chromosome 11. A similar conclusion was reached by Miyake et al.[39] Since there are reports of linkage of both schizophrenia and manic depression to chromosome 11, it has been suggested that 5HT3R-A is a possible candidate gene for these disorders.[54] Further gene mapping and sequencing studies will be required to test this possibility.

Comparison of the 5HT3R-A cDNA nucleotide, and predicted amino acid sequence with those of other Cys-loop ion channels indicates that this receptor is most highly related to the nicotinic receptor family.[30] A phylogenetic analysis[58] supports this, suggesting that the nACh and 5HT3 receptors diverged only after the evolution of distinct cation and anion channels. By contrast, the emergence of multiple nACh receptor subunits, presumably by gene duplication, occurred much later. Thus the nACh and 5HT3 receptors may have evolved from a common ancestral cation conducting ion-channel. This close evolutionary relationship is also reflected in the organization of the respective genes with the positions of introns 2, 3, and 4 within the 5HT3R-A gene completely conserved amongst all characterized nAChR genes.[33,54] The 5HT3R-A gene would appear to be most closely related to the chick nAChα7 subunit since the position of intron 7 is also conserved between these two genes. Interestingly, this subunit, and the closely related α8 nAChR, share with 5HT3R-A the ability to form functional homomeric ligand-gated ion channels when expressed heterologously (section 7.6).

7.5 TEMPORAL AND SPATIAL EXPRESSION OF 5-HT3R-A

The gross distribution 5-HT3R-A within the central nervous system and periphery has been examined utilizing RT PCR[30] or Northern blot[39] analysis. In the mouse, transcripts were detected in cerebral cortex, midbrain, brain stem, spinal cord and heart.[30] In human brain, transcripts were detected in the amygdala, caudate nucleus and hippocampus, while in the periphery hydridization occurred to RNA isolated from small intestine, colon, spleen, thymus and prostate.[39] Neither rat nor human heart demonstrated 5-HT3R-A transcripts.[39]

A more detailed examination of the spatial expression patterns of 5HT3R-A mRNA in sections of adult, female, mouse brain[59] and adult rat brain[60] has been performed using the technique of in situ·hybridization. Consistent with the distribution of 5-HT3 receptor sites in mouse and rat brain defined by autoradiographic mapping,[14,61,62-64] labelled cells were detected in mouse forebrain regions including the cerebral cortex, the hippocampal formation, amygdaloid complex, olfactory bulb and hypothalamus. In the rat, labelling was additionally detected in limbic structures including the medial habenula, medial septum, diagonal band of Broca, nucleus basalis of Meynert, nucleus accumbens and olfactory tubercle. Autoradiographic studies have detected 5-HT3 receptors in only some of these structures (e.g. septum, nucleus accumbens).[62,63] Conversely, in rat brain functional studies suggest the presence of 5-HT3 receptors in forebrain areas such as the striatum[65] where hybridization has not been reported. However, 5-HT3R-A mRNA has been amplified from the striatum using RT-PCR.[53] A number of factors might, in principle, contribute to such mismatches. These might include a possible pre-synaptic location of the 5-HT3 receptor (see below), inherent differences in the sensitivities of the techniques employed, species differences in 5-HT3 receptor distribution, and conceivably, the existence of mRNAs encoding novel 5-HT3 receptor subunits that are not recognized by the riboprobes utilized in the above studies.

Within mouse cortex,[59] hybridizing cells were found most frequently in superficial layers and occurred at highest density in allocortical (i.e. entorhinal, cingulate and piriform cortices), as opposed to neocortical regions, a distribution reminiscent of that encountered in autoradiographic studies of rat forebrain.[62-64] Expression of the 5-HT3R-A subunit within the cerebral cortex has additionally been detected with antibodies raised against a synthetic peptide representing a stretch of amino acids located within the putative large intracellular loop of the r5-HT3R-A.[44] Interestingly, immunocytochemical characterization of neocortical neurones expressing the 5-HT3R-A suggests a population of these to be GABAergic.[66]

In mouse hippocampus,[59] strong hybridization was detected throughout, but particularly in posteroventral regions. In the CA1 area, a prominent signal was obtained within the stratum lacunosum moleculare where 5-HT3 receptors have also been detected by autoradiography.[67] Low expression was detected in pyramidal neurons where autoradiographic studies have indicated the presence of 5HT3 receptors.[67] It was suggested that this apparent discrepancy could result from the presence of presynaptic 5-HT3 receptors on terminals from neurons located in other layers. Indeed, electrophysiological studies indicate that 5-HT3 receptors mediate excitation of GABAergic interneurones located within the stratum lacunosum moleculare of CA1,[68] an effect that results in an increased frequency of GABA-mediated spontaneous inhibitory postsynaptic potentials (sIPSPs) recorded from pyramidal neurones.[69] The pyramidal neurones themselves, however, are not directly affected by 5-HT3 receptor agonists[69,70] and lesioning studies suggest that these cells do not express 5-HT3 receptors.[63] The expression pattern within the CA3 field of the hippocampus (highest expression in the stratum radiatum and stratum lucidum) is also consistent with 5HT3R-A mRNA expression being largely in interneurons.[59] In the dentate gyrus, hydridizing cells are detected at the hilar edge of the granule cell layer,[59] a result consistent with electrophysiological studies demonstrating 5-HT3 receptor mediated excitation of basket cells[71] and a subsequent increase in sIPSPs recorded from granule cells.[72]

Autoradiographic studies have revealed discrete areas within the brainstem of several species to express a high density of 5-HT3 receptors.[14,15,61,63,64,67,73,74] Accordingly, in situ hydridization applied to mouse brainstem slices[59] indicates the presence of 5-HT3R-A transcripts in neurones of the nucleus of the spinal tract of the trigeminal nerve which possesses 5-HT3 binding sites.[14] However, hybridization was also observed in the nuclei of the trochlear and facial nerves,[59] where 5-HT3 recognition sites have been neither detected by autoradiography, nor, in the case of facial motoneurones, by electrophysiological studies.[75] Conversely, areas of the dorsal vagal complex (i.e. dorsal motor nucleus of the vagus nerve, nucleus of the solitary tract, area postrema) containing a high density of 5-HT3 binding sites, did not yield hybridizing cells. The latter discrepancy can most probably be attributed to many of the 5-HT3 recognition sites within the dorsal vagal complex area residing upon the presynaptic terminals of vagal afferent fibers arising from somata located within the nodose ganglion.[63,73,74]

In slices of spinal cord[59,60] obtained from both mouse and rat, hybridizing cells were detected in the superficial layers of the dorsal horn. In the latter species, a signal was also detected in the ventral horn. In the rat, 5-HT3 binding sites appear to be restricted to the superficial layers of the dorsal horn with the majority of the labelling representing receptors located upon the central terminals of primary sensory afferents, particularly C-fibers,[63] with the remainder perhaps corresponding to receptors located upon GABAergic and glycinergic spinal interneurones.[76] In agreement with a presynaptic location of the receptor upon sensory afferents within the cord, a subpopulation of dorsal root ganglion neurones contain 5-HT3R-A transcripts.[59] Elsewhere within the peripheral nervous system, 5-HT3R-A mRNA has been detected in both the myenteric and the submucous plexus of the duodenum, jejunum and ileum of the rat.[77] Corresponding binding sites in this species are of low density and their detection has been restricted to the terminal portions of the small intestine.[78]

The expression of 5HT3R-A mRNA during mouse and rat embryogenesis has also been investigated and in both species there is widespread expression during the development of both neuronal and non-neuronal tissues.[79,80] In the mouse, expression was detected within the cranial sensory ganglia at day 9.5, which was the earliest time point examined.[79] By E10.5, expression was also evident within the nodose and petrosal ganglia, as well as the trigeminal ganglion. Expression within the dorsal root ganglion appeared slightly later (E12.5) and by E14.5 there was also expression at all levels of the gut as well as in the developing sympathoadrenal system. Within the CNS, mRNA was first detected in the medial ganglionic eminence, a structure from which the amygdala develops. At E14.5 transcripts were also observed in the developing cortex and hippocampus and within regions of the hypothalamus, pituitary and substantia nigra. A similar pattern of expression was observed in the rat, with expression first detected in the cranial sensory ganglia at E10.[80] These studies also reported the presence of 5HT3R-A transcripts in developing non-neuronal tissues including the axial skeleton, craniofacial regions, olfactory epithelium, lung, and developing tooth bud and genitalia. In the adult, the expression of 5HT3 receptors is thought to be restricted to neurons.

Miquel et al[53] have utilized quantitative PCR to investigate the relative levels of expression of mRNAs encoding the 5-HT3R-A_S and 5-HT3R-A_L splice variants during the development of the rat nervous system. Their data suggest the ratio of 'long' to 'short' forms of 5-HT3R-A mRNA in cerebral cortex, hippocampus and striatum to decrease during late embryonic development (i.e. E17 to E20) with a further reduction in the relative expression of the 'long' form occurring into adulthood. A particularly marked trend was observed for the nodose ganglion, where 5-HT3R-A_L was the predominant isoform at E20 (75% of 5-HT3R-A mRNA) but represented only minor fraction

(10%) at adulthood. A qualitatively similar phenomenon was noted to occur in the maturing superior cervical ganglion. These interesting findings suggest a developmental influence upon the mechanism that governs the choice of splice acceptor sites. However, as discussed in section 7.6, the functional significance of the alternatively spliced forms of 5-HT3R-A is at present unclear.

The pattern of 5HT3R-A expression during rodent embryogenesis is particularly intriguing as it raises the possibility that 5-HT3 receptors could mediate some of the effects of 5-HT during the development of both neuronal and non-neuronal tissues. However, studies which detect only the presence of a given mRNA have to be interpreted with some caution and possible relationships to function remain speculative. Since ultimately genetic models should help to clarify the role of a given gene product, we are in the process of creating transgenic mice lacking the 5HT3R-A subunit.

7.6 FUNCTIONAL PROPERTIES OF RECOMBINANT 5-HT3 RECEPTORS

The ability of 5-HT3 receptor cDNAs, isolated from different species (see section 7.2), or in vitro transcripts derived from them, to drive the heterologous expression of functional 5-HT3 receptors has been assessed in both *Xenopus laevis* oocytes and mammalian cells. Cloned 5-HT3 receptor subunits isolated from mouse,[30,32,33,50,51,81-84] rat[36] and human[38,39,85] assemble efficiently, in both systems, to form presumably pentameric receptor complexes,[42] exhibiting many of the properties associated with their native counterparts. With the exception of some discrepancies in agonist pharmacology (see below) the properties of the 'long' and 'short' m5-HT3R-A subunits appear to be identical.[33,83] Thus, the properties of m5-HT3R-A will be dealt with collectively in the following discussion.

In cells voltage-clamped at negative holding potentials, recombinant 5-HT3 receptors mediate a transient inward current response which desensitizes in the continued presence of agonist. The concentrations of 5-HT that evoke a half maximal inward current response (i.e. EC_{50} or K_{app}) in cells expressing either m5-HT3R-A_S or m5-HT3R-A_L[30,32,33,50,51,81,83,84] fall within a narrow range of values (1-4 μM) that corresponds to those reported for the 5-HT3 receptor expressed by murine neuronal cell lines[86-88] and mouse superior cervical ganglion neurones[81] (see Table 7.2). Similarly, the EC_{50} for 5-HT acting upon r5-HT3R-A_S[37] expressed in oocytes (≈1.5 μM; I Mair, unpublished observations) approximates to values reported for 5-HT3 receptors of rat PC12 cells[57] and rat nodose ganglion neurones.[89] Electrophysiological studies performed on native 5-HT3 receptors report Hill coefficients for 5-HT3 receptor mediated responses ranging between 1.2 and 3 (reviewed in ref. 26), suggesting that at least two molecules of 5-HT are required to activate native 5-HT3 receptors. Responses mediated

Table 7.2. Properties of m5-HT_3R-A splice variants expressed in* Xenopus laevis *oocytes or mammalian cell lines in comparision to 5-HT_3 receptors native to murine neuroblastoma cell lines.

	m5-HT_3R-A_S			**m5-HT_3R-A_L**			**5-HT_3 N1E-115**		
Biophysical properties									
Reversal potential (mV)	+2.3[a]			-2.2[b]			-2.7[a]		
Single channel conductance (pS)	0.36[a]			0.42[b]			0.55[a] 0.38[c]		
Pharmacological properties	pEC_{50}/ pIC_{50}[d]	nH[d]	α[d]	pEC_{50}/ pIC_{50}[d]	nH[d]	α[d]	pEC_{50}/ pIC_{50}[e]	nH[e]	α[e]
Agonists									
5-HT	5.63	2.2	1.00	6.00	2.6	1.00	5.38	2.3	1.0
2-Methyl-5-HT	4.83	2.2	0.09	4.89	1.8	0.63	4.76	1.2	0.2
m-Chlorophenylbiguanide	6.06	1.7	0.84	6.40	1.5	0.87	5.85	1.5	1.0
Phenylbiguanide	4.52	1.7	0.76	4.74	1.8	0.82	ND	ND	ND
Antagonists									
Granisetron	9.85	-	-	9.80	-	-	9.90	-	-
BRL 46470	9.85	-	-	9.82	-	-	ND		
Ondansetron	8.92	-	-	9.05	-	-	9.60	-	-
(+)-Tubocurarine	8.85	-	-	8.40	-	-	9.09	-	-
Bemesetron	8.51[a]	-	-	ND	-	-	8.46[a]	-	-
Metoclopramide	7.14	-	-	7.20	-	-	7.62	-	-
Cocaine	5.63	-	-	5.47	-	-	5.14	-	-

Notes: Data obtained under-voltage clamp conditions for 5-HT_3 receptors native to murine N1E-115 neuroblastoma cells (refs. a, b and e) or for 5-HT_3R-A splice variants heterologously expressed in either Xenopus laevis *oocytes (ref. d) or mammalian cells (refs. a, b and c). Data were collated from (a) Hussy et al,[81] (b) Gill et al,[82] (c) Lambert et al,[23] (d) Downie et al[83] and (e) Lambert et al.[28]*

by all presently identified species homologs of 5-HT3R-A demonstate Hill coefficients greater than unity,[30,32,38,39,81,83] suggesting that activation of the homo-oligomeric receptors is also a cooperative process.

5-HT3 receptors endogenous to a variety of neurones and neuronal cell lines are cation selective ion channels that discriminate poorly between monovalent metals.[23,24,29,57,90-93] The reversal potential of the agonist evoked responses mediated by both the murine and human cloned 5-HT3 receptors are close to 0 mV,[30,50,51,81,82,85] consistent with the activation of a channel that conducts monovalent cations non-selectively. In addition, limited ion substitution experiments performed on mouse[30,82] and human[85] recombinant 5-HT3 receptors in which external sodium ions were replaced by either Tris,[30] or N-methyl-D-glucamine,[82,85] or the intracellular concentration of monovalent metals was reduced by replacement with the poorly permeant tetraethylammonium ion[82] (AM Brown, unpublished observations), produced displacements of the reversal potential indicative of a cation selective ion channel which discriminates poorly between monovalent cations.

Although the inward current response mediated by 5-HT3 receptors can be attributed mainly to the transmembrane flux of sodium and potassium ions, native 5-HT3 receptors may also display a significant permeability to certain divalent cations.[90,92] The influx of calcium ions through 5-HT3 receptors native to murine N1E-115 cells, as well as m5-HT3R-A expressed in HEK-293 cells, can be visualized by calcium imaging in cells pre-loaded with fura 2.[94] Thus, 5-HT and the selective 5-HT3 receptor agonist *m*-chlorophenylbiguanide (*m*-CPBG), concentration-dependently increase intracellular levels of calcium, a response which is blocked by the 5-HT3 receptor antagonist ondansetron. Furthermore, the elevation of cytosolic calcium concentrations elicited by 5-HT3 receptor agonists is abolished by the removal of extracellular calcium, and is independent of the status of intracellular calcium stores.[94] By contrast, evidence also exists which suggests native[91] and cloned[95] 5-HT3 receptors to be impermeable to calcium. Furthermore, although it is well established that the divalent cations Ca^{2+} and Mg^{2+} when present extracellularly at physiological concentrations can exert a voltage-independent blockade of 5-HT3 receptors native to neuronal cell lines[90,91,96] and peripheral neurones,[93] conflicting data exist concerning the inhibition of 5-HT3R-A by these divalent metals. In several studies of mouse[30,51] and rat[36] 5-HT3R-A expressed in oocytes, Ca^{2+} and Mg^{2+} exerted a voltage-dependent blockade of inward current responses to 5-HT manifested as the introduction of a region of negative slope conductance into the 5-HT current-voltage relationship. Such data are consistent with blockade of the 5-HT3R-A channel by divalent cations, a phenomenon which been interpreted to hint at the co-expression of other subunits, which confer protection from this effect, in the 5-HT3 receptor complex native to neuronal cell lines and peripheral neurones.[30] However, there is no evidence that Ca^{2+} or Mg^{2+}

exert a voltage-dependent blockade of either mouse[81,82] or human (AM Brown, unpublished observations) recombinant receptors expressed in mammalian cells. In the latter studies, inwardly rectifying current voltage relationships similar to those reported for native 5-HT3 receptors were observed. Additionally, although Yakel et al[50] found Ca^{2+} to accelerate the desensitization of currents mediated by m5-HT3R-A expressed in oocytes in a voltage-dependent manner, no clear effect upon the peak of the response was discerned. Most recently, currents mediated by 5-HT3 receptors endogenous to basket cells of the rat dentate gyrus have been reported to exhibit voltage-dependent block by Ca^{2+} and Mg^{2+}.[71] In view of the above discrepancies, it would be premature to conclude that the mode of action of Ca^{2+} and Mg^{2+} upon recombinant and native 5-HT3 receptors is fundamentally different.

The group IIb divalent metal ions Zn^{2+} and Cd^{2+} produce voltage-dependent inhibition of 5-HT3 receptor mediated responses in NCB-20 cells.[97] In addition, a variety of metals including Hg^{2+}, Pb^{2+}, Zn^{2+} and La^{3+}, but not Cd^{2+}, are reported to suppress 5-HT3 receptor-evoked currents in N1E-115 neuroblastoma cells.[98] For m5-HT3R-A expressed stably in HEK-293 cells,[82] and transiently in oocytes (CH Gill and D Belelli, unpublished observations), Zn^{2+} exhibits a biphasic effect, inhibiting 5-HT evoked responses at high (>10 μM) concentrations but potentiating the response at low (< 10 μM) concentrations. While that observation may suggest the subunit complement of the 5-HT3 receptor native to NCB-20 cells to be different to the homo-oligomeric m5-HT3R-A, it should be noted that Zn^{2+} produces only a simple monophasic inhibition of the human 5-HT3R-A subunit expressed in *Xenopus* oocytes.[99]

Single channel conductances associated with 5-HT3 receptors vary widely between preparations (reviewed in refs 25, 26, 28 and 29). In the neuronal clonal cell lines N1E-115[23,81] and N18,[90] single channel conductances of less than 1 pS have been inferred from fluctuation analysis of macroscopic currents recorded under voltage-clamp using the whole-cell recording mode of the patch clamp technique and quasi-physiological concentrations of permeant ions. Under similar recording conditions, single channel conductance values of 0.4-0.6 pS have been determined for the splice variants of m5-HT3R-A expressed in HEK-293 cells, or RIN cells by noise analysis of 5-HT-evoked currents recorded from whole cells and outside-out membrane patches[33,81,82] (see Table 7.2). Ongoing studies in our laboratory suggest the single channel conductance of the h5-HT3R-A subunit stably expressed in HEK-293 cells to have a similar, sub-picoSiemen, conductance (AM Brown, unpublished observations). However, it has recently been reported that the use of modified recording solutions which enhance the inwardly directed electrochemical gradient for Na^+ ions allows the resolution of discrete 5-HT3 receptor-gated single channel currents corresponding to a conductance of ≅ 5 pS in outside-out patches excised from N1E-115 cells.[100] In cell attached patches of the same cell line, van Hooft and Vijverberg[101] report 5-HT3 receptors to mediate the opening of a channel that demonstrates multiple conductance states of 27, 18, 11 and 6 pS. Those investigators additionally noted the relative frequency of the highest conductance state to be decreased by the protein kinase inhibitor staurosporin, whereas phorbol ester mediated activation of protein kinase C produced the opposite effect.[101] Similar experiments remain to be performed upon recombinant 5-HT3 receptors, but it is worthy of note that staurosporin inhibits 5-HT3 receptor mediated macroscopic currents in rat dorsal root ganglion neurones[102] and activators of protein kinase C augment the amplitude of 5-HT-evoked responses recorded from oocytes expressing m5-HT3R-A.[103] In the latter study, staurosporin, but not H-89 or genistein which inhibit protein kinase A and tyrosine kinase, respectively, depressed the response to 5-HT. Interestingly, the primary amino acid sequence of m5-HT3R-A does not contain a recognized consensus sequence for phosphorylation by protein kinase C.[30]

Single channel conductances associated with 5-HT3 receptors expressed by neurones of guinea-pig submucous plexus (9 and 15 pS)[24] and coeliac ganglion (10 pS)[104], rabbit nodose ganglion (19 pS)[93], rat superior cervical ganglion (11 pS)[92], and mouse hippocampus[105] and superior cervical ganglion (10 pS)[81] are all substantially larger than those reported for neuronal cell lines, or m5-HT3R-A (when standard recording solutions are employed). Cell-specific post-translational modifications of a common receptor subunit might conceivably contribute to these discrepancies; the phosphorylation status of the receptor perhaps representing one pertinent variable.[101] It has also been suggested from studies performed on the hybridoma cell line NG 108-15, that the conductance of the 5-HT3 receptor complex varies between differentiated and undifferentiated cells.[106] Although the ratio of 5-HT3R-A_S and 5-HT3R-A_L transcripts in NG 108-15 cells is also influenced by cellular differentiation,[55] this cannot explain the change in channel conductance because homo-oligomeric receptors composed of 5-HT3R-A_S or 5-HT3R-A_L subunits exhibit identical biophysical properties.[33,81,82] It is possible that subunits additional to 5-HT3R-A are contained within receptor-channel complexes that demonstrate a substantial single channel conductance. In this respect it is interesting to note that fluctuation analysis of 5-HT3 receptor mediated whole-cell currents recorded from mouse and rat superior cervical ganglion neurones suggests that in addition to those channels that can be resolved directly, 5-HT3 receptors of low single channel conductance co-exist within the same cell.

The agonist and antagonist pharmacology observed for all of the available cloned 5-HT3 receptors mimics faithfully that reported for native 5-HT3 receptors.[30,32,38,81-83,85] Data obtained from electrophysiological studies in which either m5-HT3R-A_S, or m5-HT3R-A_L were expressed in oocytes, or mammalian cells are summarized in Table 7.2

which compares the properties of the recombinant receptors with those of 5-HT3 receptors endogenous to a representative murine neuronal cell line (i.e. N1E-115). Such a quantitative comparison reveals the potencies of all agonist and antagonist compounds examined to date to be invariant across the receptors. However, a subtle difference in receptor properties is suggested by the data obtained with 2-methyl-5-HT, which evoked peak current responses amounting to 9% and 63% of those elicited by a saturating concentration of 5-HT when acting upon 5-HT3R-A_S and 5-HT3R-A_L homo-oligomers, respectively.[83] Unfortunately, it is difficult to assess the significance of this finding, because although qualitatively similar results have been obtained with 2-methyl-5-HT in experiments conducted upon the two splice variants expressed in HEK 293 cells,[107] a much less marked difference in the efficacy of 2-methyl-5-HT was detected when the recombinant receptors were studied in RIN cells.[33] Additionally, it is not immediately obvious how the absence or presence of six amino acids in the large cytoplasmic loop (see section 7.3) selectively influences the efficacy of 2-methyl-5-HT. Indeed, at h5-HT3R-A_S which also lacks the hexapeptide, 2-methyl-5-HT evokes maximal current responses only slightly less,[38] or equal to,[39] those elicited by a saturating concentration of 5-HT.

Numerous studies employing functional assays[9,108-110] and radio-ligand binding techniques[64,111-113] have unequivocally demonstrated that the pharmacological profile of the 5-HT3 receptor varies between species (reviewed in refs. 25, 26 and 28). By comparison, the limited pharmacological evidence that supports the existence of intraspecies subtypes of 5-HT3 receptor[114,115] has proved difficult to reproduce.[116] Interspecies heterogeneity in 5-HT3 receptor properties is most clearly exemplified by the receptor expressed by guinea-pig tissues which demonstrates a generalized reduction in affinity towards antagonists in comparision to the receptor expressed other species.[108,109,112] In addition, the agonist 1-phenylbiguanide (PBG) lacks efficacy at the guinea-pig receptor.[108,109] Although not as extensively documented, clear distinctions can be made between 5-HT3 receptors present in human, mouse, rat, and rabbit tissues using compounds that include (+)-tubocurarine[109,117] and the arylbiguanides PBG and *m*-chlorophenylbiguanide (*m*-CPBG).[111,112] Table 7.3 provides a summary of the pharmacological profile of human, rat and mouse recombinant 5-HT3 receptors determined in radioligand binding assays applied to membrane homogenates derived from mammalian cells stably, or transiently, expressing the receptor. For comparision, the table also lists binding parameters for 5-HT3 receptors endogenous to human, rat and mouse brain. Inspection of the table emphasizes how closely the recombinant receptors reproduce the pharmacological properties of their endogenous counterparts. To take the most striking example, the affinity (expressed as pK_i) of (+)-tubocurarine for m5-HT3R-A_S and 5-HT3 receptors native to mouse brain is approximately 200-fold higher than that reported for

Table 7.3. Pharmacological profiles of mouse, rat and human recombinant 5-HT_3R-A in comparision to native receptors from the corresponding species

Drug	pKi										
	m5-HT_3R-A_L[a,b] or m5-HT_3R-A_S[c,d]		Mouse Cortex		r5-HT_3R-A_S	Rat Cortex		h5-HT_3R-A_S		Human Putamen[i] or Amygdala[j]	
Agonists											
*m*CPBG	7.5[a]	6.64[d]	7.75[e]	7.0[f]	8.98[g]	9.22[h]	-	6.65[g]	7.18[i]	7.00[j]	-
5-HT	6.57[b]	6.26[c]	-	6.4[f]	7.39[g]	7.94[h]	6.6[f]	6.47[g]	6.91[i]	6.80[j]	6.89[k]
2-Me-5-HT	6.1[a]	5.54[c]	6.32[e]	5.6[f]	7.22[g]	7.68[h]	6.8[f]	6.18[g]	6.64[i]	6.22[j]	-
PBG	6.2[a]	5.54[c]	6.27[e]	5.5[f]	-	-	7.3[f]	-	4.61[i]	5.33[j]	-
Antagonists											
YM060	10.1[a]	-	-	10.2[f]	10.98[g]	11.47[h]	10.6[f]	10.21[g]	-	-	-
YM114	8.0[a]	-	-	7.8[f]	9.95[g]	10.76[h]	7.9[f]	9.93[g]	-	-	-
Granisetron	-	8.67[d]	8.64[e]	-	8.92[g]	9.55[h]	-	8.60[g]	8.85[i]	8.03[j]	-
Tropisetron	8.69[b]	8.42[c]	-	8.9[f]	8.90[g]	9.74[h]	8.9[f]	8.48[g]	-	7.89[j]	8.16[k]
Ondansetron	-	7.87[d]	8.32[e]	8.1[f]	8.52[g]	8.89[h]	8.1[f]	8.34[g]	8.31[i]	7.76[j]	8.42[k]
Bemesetron	8.06[b]	-	7.85[e]	-	-	-	7.7[f]	-	-	-	7.92[k]
Metoclopramide	-	-	6.68[e]	6.6[f]	-	-	6.2[f]	-	6.44[i]	5.87[j]	6.25[k]
Cocaine	>5.0[a]	-	5.29[e]	5.3[f]	-	-	5.4[f]	-	5.38[i]	4.79[j]	-
(+)-Tubocurarine	8.06[b]	7.59[d]	7.75[e]	7.3[f]	-	-	6.5f[e]	-	4.79[i]	4.41[j]	-

Notes: Data refer to species homologs of 5-HT_3R-A stably, or transiently expressed, in mammalian cells and characterized in radioligand binding assays performed upon membrane homogenates. All values are, with the exception of data for m5-HT_3R-A_S, the pK_i values obtained in competition studies between the listed drug and radiolabelled 5-HT_3 receptor antagonists. Data for m5-HT_3R-A_S identified by the superscript 'c' are listed as untransformed pIC_{50} values. Data were compiled from the following sources (radioligand in parenthesis): (a) Bonhaus et al[131] ([^{3}H]RS-42358); (b) Maricq et al[30] ([^{3}H]GR63650); (c) Downie et al[122] ([^{3}H]granisetron); (d) Hope et al[132] ([^{3}H]granisetron); (e) Perren et al[116] ([^{3}H]granisetron); (f) Bonhaus et al[114] ([^{3}H]RS-42358-179); (g) Miyake et al[39] ([^{3}H]YM060); (h) Akuzawa et al[133] ([^{3}H]YM060); (i) Hope et al[134] ([^{3}H]granisetron); (j) Bufton et al[135] ([^{3}H]granisetron) and (k) Abi-Dargham et al[136] ([^{3}H]LY278584).

h5-HT3R-A_S and 5-HT3 receptors endogenous to human brain. Functional studies are in qualitative agreement with this finding. (+)-Tubocurarine inhibits electrical responses mediated by m5-HT3R-A[32,81-83] and receptors native to mouse neurones[109,110,117,118] at low nanomolar concentrations, but is reported to exhibit a potency some 100- and 1800-fold lower at rat[36] and human[38,39] cloned receptors, respectively. A second clear example of interspecies heterogeneity in the properties of both recombinant and native 5-HT3 receptors is provided by the arylbiguanides. The affinity of *m*-CPBG for r5-HT3R-A_S and the receptor of rat cortex is approximately 100-fold higher than that found for h5-HT3R-A_S or the receptor native to human brain. Qualitatively similar findings have been made in functional studies examining the agonist potency of *m*-CPBG at rat and human isoforms of 5-HT3R-A expressed in oocytes (I Mair, unpublished observation). The data obtained with arylbiguanidies are the most pronounced examples of a trend which suggests that agonist, but not antagonist, compounds exhibit higher affinities for rat 5-HT3 receptors in comparison to human and mouse isoforms (Table 7.3).

5-HT3 receptors endogenous to murine neuronal cell lines and rat peripheral neurones are subject to positive allosteric regulation by alcohols that include ethanol and trichloroethanol.[118-120] Recent studies have demonstrated similar regulation of m5-HT3R-A expressed in oocytes and mammalian cells, and have extended the range of modulators to include butanol, isopentanol and the volatile general anaesthetic agents isoflurane, halothane and 1,2,2-trifluorocyclobutane.[121-123] Potentiation of 5-HT evoked currents by alcohols occurs at either splice variant of m5-HT3R-A,[122] and exhibits concentration-dependency over a clinically relevant range. Based upon the dextral displacement of the 5-HT concentration effect relationship in the presence of trichloroethanol,[122] and diminished degree of potentiation by alcohols when high concentrations of 5-HT are employed,[122,123] it is plausible that an increase in receptor affinity for the agonist contributes to the modulatory effect. Indeed, in binding assays performed on HEK 293 cell membrane homogenates containing the m5-HT3R-A_S, trichloroethanol increased the apparent affinity of agonist binding, but had no effect upon the binding of antagonists.[122] However, other mechanisms must also be operative, because the maximal current response elicited by saturating concentrations of partial agonists,[122] but not 5-HT,[122,123] was dramatically enhanced by trichloroethanol. It is currently not known whether the volatile agents share this property. It is tempting to speculate that enhancement of 5-HT3 receptor function by volatile agents may contribute to the phenomenon of post-operative nausea and vomiting. However, caution is necessary since evidence also exists that several intravenous anaesthetic agents, including some that clearly provoke emesis, suppress 5-HT3 receptor-mediated cation fluxes in N1E-115 neuroblastoma cells.[124]

In summary, the biophysical and pharmacological properties of species homologs of recombinant 5-HT3 receptors resemble, to a remarkable degree, those associated with native 5-HT3 receptors. It must be concluded that at present there is little evidence for, or reason to suggest, the existence of additional 5-HT3 receptor subunits. This situation may, of course, change as more detailed comparisons of native and recombinant 5-HT3 receptors are made.

7.7 STRUCTURE FUNCTION RELATIONSHIPS

Although the amino acid sequence of the 5-HT3R-A subunit has been established for a number of species (see section 7.2), the present understanding of the relationship between primary structure and receptor function is presently rudimentary. Biochemical studies in which membrane homogenates prepared from NG108-15 cells were incubated with protein modifying agents have provided limited information concerning amino acid residues involved in ligand binding.[125] Treatment of membrane homogenates with N-bromosuccinimide, which selectively oxidizes tryptophan residues, inhibited the binding of the antagonist [^{3}H] zacopride, decreasing B_{max} in the absence of any effect upon K_D. In addition, a smaller contribution of histidine, tyrosine and arginine has been suggested following modification by diethylpyrocarbonate (DEPC), tetranitromethane and 2,3-butanedione, respectively.[125] The reduction in the number of [^{3}H] zacopride binding sites by protein modification could be prevented by preincubation with the 5-HT3 receptor antagonists granisetron, ondansetron and bemesetron and the agonist PBG, but not by *m*-CPBG, zacopride or (s)-zacopride. Ligand binding mechanisms at the 5-HT3 receptor may thus vary between ligands.

By analogy to the nicotinic receptor,[40,41] the 5-HT3 receptor is envisaged as being subdivided into two functional domains; an extracellular ligand-binding domain at the amino terminus, and a carboxyl terminal region which contains transmembrane and cytoplasmic sequences (see section 7.2). This proposed organization is supported by evidence from a study of the properties of a chimera formed between the 5-HT3R-A subunit and the closely related chick nicotinic AChR α7 subunit.[51] Heterologous expression of the chimaeric receptor subunit, in which the putative extracellular domain of the 5-HT3R-A subunit was replaced by the corresponding region of the chick α7 subunit, resulted in the synthesis of homo-oligomeric receptors with pharmacological and allosteric properties reminiscent of the nAChR. By contrast, the biophysical properties of the ion channel integral to the recombinant receptor were characteristic of the murine 5-HT3 receptor.[51]

The assignment of the ligand binding domain to an extracellularly located 'module'[40] can be further tested by exploiting species differences in 5-HT3 receptor pharmacology. In a recently engineered chimaeric construct, the entire extracellular N-terminal domain of the h5-HT3R-A_S (which binds (+)-tubocurarine with low affinity–see section 6) was replaced by the corresponding domain of the m5-HT3R-A (for which (+) tubocurarine demonstrates high affinity). The chimera was functional upon expression in oocytes and, as expected from the similar potency of 5-HT at m5-HT3R-A_S[32,83] and h5-HT3R-A_S[38,39] subunits, the EC_{50} for activation by 5-HT was unchanged (AG Hope and D Belelli, unpublished observations). In marked contrast, the IC_{50} for (+)-tubocurarine decreased dramatically to a value similar to that found for wild-type m5-HT3R-A_S (see Fig. 7.4).

Two recent preliminary reports have identified specific amino acids within the extracellular module of m5-HT3R-A which impact upon ligand binding. From sequence comparisions between m5-HT3R-A and nicotinic receptor subunits, Schulte et al[126] identified a specific tryptophan residue (W66 in m5-HT3R-A) as a potential structural determinant of high affinity (+)-tubocurarine binding. This residue corresponds to an homologous tryptophan located within the 'D-loop' of the α and β subunits of the nicotinic receptor of *Torpedo* which has been implicated in (+)-tubocurarine binding.[40] Mutation of W66 to alanine (A) resulted in the loss of specific [^{3}H]-granisetron binding, whereas replacement by phenylalanine (F) produced a 4-fold decrease in the affinity of the radioligand. The W66F mutant also displayed slightly reduced apparent affinities (K_i) for 5-HT and bemesetron, but the most pronounced effect was an approximately 14-fold decrease in the affinity of (+)-tubocurarine binding.[126] W66 is conserved at an homolgous position across all species homologs of 5-HT3R-A. Other, non-conserved, residues must additionally contribute to the (+)-tubocurarine binding site in order to account for the marked species-dependent variation in (+)-tubocurarine binding affinity. Indeed, recent studies by our own group using oligonucleotide-mediated mutagenesis implicate residues within a region homologous to loop C of nicotinic receptor subunits (AG Hope and D Belelli, unpublished observations).

Fig. 7.4. (opposite page) The differential sensitivity of h5-HT$_3$R-A_S and m5-HT$_3$R-A_S to antagonism by (+)-tubocurarine resides within the extracellular N-terminal domain. Left: schematic representations of wild-type m5-HT$_3$R-A_S (top), wild-type h5-HT$_3$R-A_S (bottom) and a chimaeric construct (middle) which is composed from the entire extracellular N-terminal 'module' of m5-HT$_3$R-A_S (clear) and the transmembrane and cytoplasmic 'module' of h5-HT$_3$R-A_S (stippled). Positions of putative transmembrane sequences are indicated by the solid vertical bars. Right: inward current responses to an approximate EC_{50} of 5-HT (3 (M) bath applied to oocytes expressing either m5-HT$_3$R-A_S (top), h5-HT$_3$R-A_S (bottom) or the chimeric construct (middle). Note: (i) the large difference in the concentration of (+)-tubocurarine required to cause an approximately equivalent degree of block at m5-HT$_3$R-A_S vs. h5-HT$_3$R-A_S and (ii) that the sensitivity of the chimera to the blocker matches that of m5-HT$_3$R-A_S. Current responses were recorded at a holding potential of -60 mV and periods of drug applications are indicated by the horizontal bars above each trace. Electrophysiological data courtesy of Dr. D. Belelli (unpublished observations).

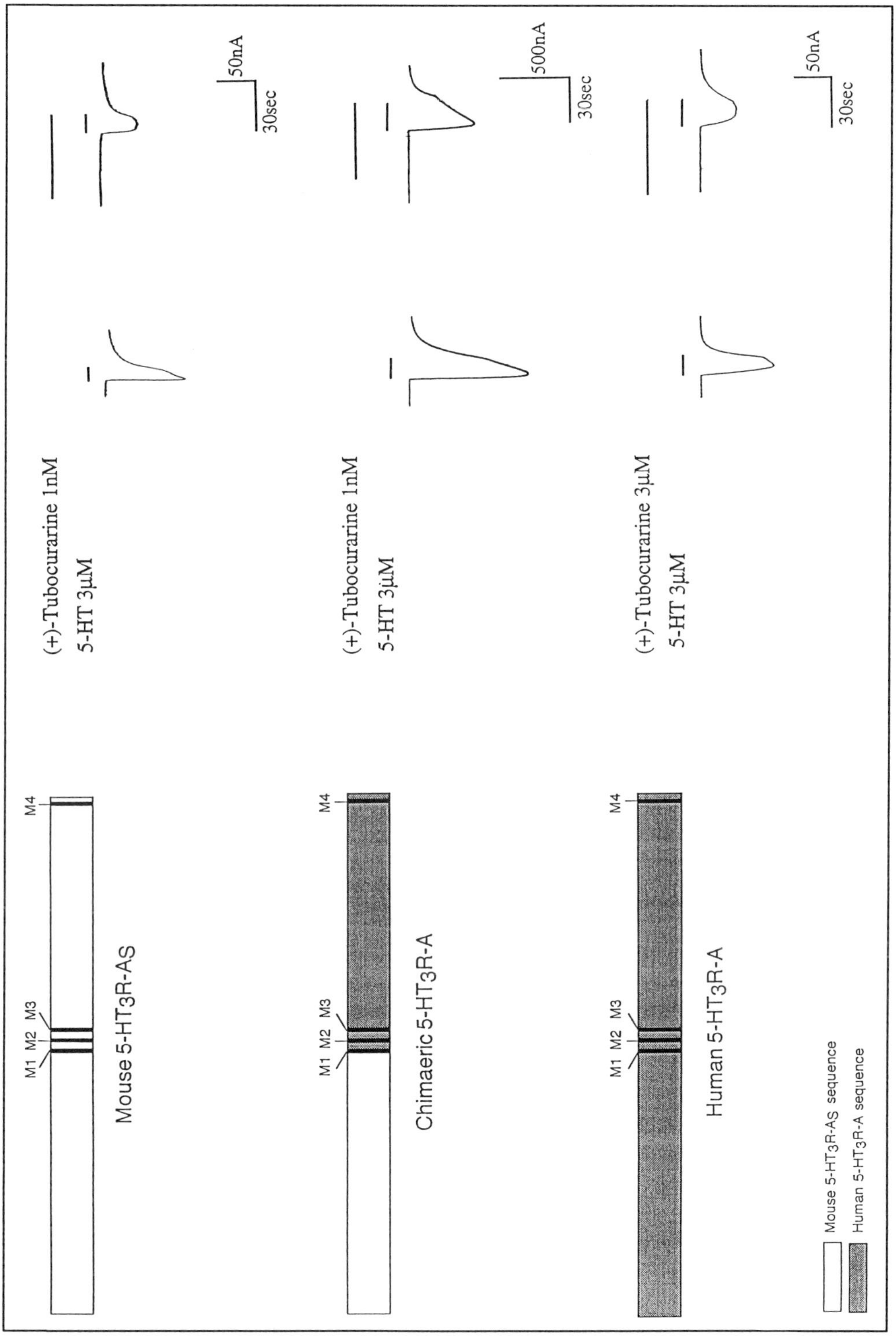
(+)-Tubocurarine 1nM
5-HT 3μM
50nA
30sec
(+)-Tubocurarine 1nM
5-HT 3μM
500nA
30sec
(+)-Tubocurarine 3μM
5-HT 3μM
50nA
30sec
M1 M2 M3 M4
Mouse 5-HT3R-AS
Chimaeric 5-HT3R-A
Human 5-HT3R-A
Mouse 5-HT3R-AS sequence
Human 5-HT3R-A sequence

Profound effects upon the binding of agonists, to what appears to be a desensitized state of m5-HT3R-A, following mutation of a glutamate residue (E106) have recently been described.[84] At the mutant E106N, the affinity of the antagonist [^{3}H]GR65630 was similar to that observed for wild-type m5-HT3R-A, but the affinity with which 5-HT and 2-methyl-5-HT competed for such binding was reduced by 30- and 17-fold, respectively. A reduction in the affinity of the antagonist renzapride was also noted, but interestingly, the K_i for *m*-CPBG was unaffected. At a second mutant, E106D, the affinities of 5-HT and 2-methyl-5-HT were more substantially depressed (132- and 246-fold, respectively), but in this case, a decrease in the affinity (14-fold) of [^{3}H]GR65630 was also noted. The K_i of *m*-CPBG was little affected and that of renzapride increased 22-fold. In parallel electrophysiological assays, the EC_{50} for 5-HT was unaffected by the E106D mutation, but increased 8-fold at the E106N mutant.[84]

By analogy to other members of the Cys-loop family of ionotropic receptors, the second transmembrane domain of 5-HT3R-A is thought to contribute to the lining of the cation selective ion channel (Fig. 7.2). Sequence alignments between nicotinic receptor subunits and 5-HT3R-A (reviewed in ref. 28) suggest that conserved acidic and polar amino acids may exert similar influences upon ion permeability and conductance within the two receptor types, but experimental verification of such predictions is currently lacking. The role of a specific leucine residue (L286 in the m5-HT3R-A_L sequence) within M2 has, however, been evaluated.[50] In chick nicotinic α7 receptor subunit, replacement of the homologous leucine by threonine results in the appearance of a novel conductance state associated with reduced desensitization, greatly increased sensitivity to acetylcholine and susceptibility to activation by compounds formerly acting as competitive and non-competitive antagonists.[127] A model to explain these findings suggests that residues comprising an equatorial leucine ring in a pentameric assembly of subunits swing into the channel lumen in the desensitized state and physically occlude it. The equivalent mutation introduced into the channel domain of m5-HT3R-A_L retards the rate of onset of desensitization and results in a modest increase (2-fold) in the potency of 5-HT.[50] The latter contrasts with the pronounced effect of the mutation upon the potency of acetylcholine as an activator muscle-type nicotinic receptors, where the magnitude of the "leftward" shift in the concentration effect relationship correlates with number of subunits into which the mutation is introduced.[128,129] Moreover, the 5-HT3R-A_L L286T mutant was not activated by antagonists.[50] The role of L286 in the function of 5-HT3R-A_L is thus somewhat different to that found for nicotinic receptors and suggests that models established from studies upon the latter should be applied to 5-HT3R-A with caution.

The presence of a positively charged lysine residue (Fig. 7.2) within the M2 segment of 5-HT3R-A is a unique feature in the Cys-loop family of receptors that might conceivably be related to the exceptionally low conductance that is characteristic of the recombinant receptor. However, there is no evidence to support a simplistic placement of the positively charged residue within the conduction pathway, where it could suppress cation translocation by electrostatic repulsion. Indeed, a projection of the sequence of the 5-HT3R-A M2 segment onto a helical wheel representation of the M2 domain of a nicotinic β-subunit[130] suggests the lysine residue to lie on the face of the helix that is opposite to the channel lumen. Maricq et al[30] hypothesize that this energetically unfavorable placement is made possible by interaction between the lysine residue and an appropriately located aspartate residue within the M1 domain. Preliminary studies in our own laboratory, in which the lysine has been mutated to glycine (the homologous residue in the chick α7 sequence), indicate a greatly reduced rate of onset of desensitization, but a more detailed characterization remains to be performed.

7.8 CONCLUSION

In conclusion, only one 5-HT3 receptor subunit, which exists as species homologs, and in mouse and rat at least, as two splice variants, has been identified to date. The recombinant homo-oligomeric complex exhibits properties remarkably similar to those of endogenous receptors. At present, it would appear that species variations in the pharmacological properties of the latter can be entirely attributed to subtle variations in 5-HT3R-A sequence. There is little evidence for intraspecies variants of 5-HT3 receptor. Whether additional, structurally distinct, 5-HT3 receptor subunits exist is obviously a matter for conjecture; large discrepancies in single channel conductance between recombinant and 5-HT3 receptors endogenous to neurones are perhaps the best indication that subunit diversity might eventually be disclosed.

In the meantime, the 5-HT3 receptor provides interest as the sole example of an ion channel activated by monoamines, as an experimental system that can be exploited to test the generality of models proposed from studies conducted upon related ionotropic receptors, and, of course, as a therapeutic target whose potential remains to be fully evaluated.

ACKNOWLEDGMENTS

The authors' experimental work upon recombinant 5-HT3 receptors is supported by the Wellcome Trust. We thank our colleagues, Dr. AM Brown, Dr D Belelli, CH Gill and I Mair, for permission to include their unpublished observations.

REFERENCES

1. Rocha e Silva M, Valle JR, Picarelli, ZP. A pharmacological analysis of the mode of action of serotonin 5-hydroxytryptamine upon the guinea-pig ileum. Br J Pharmacol 1953; 8:378-88.
2. Gaddum JH, Picarelli ZP. Two types of tryptamine receptor. Br J Pharmacol 1957; 12: 323-28.
3. Wallis DI. Neuronal 5-hydroxytryptamine receptors outside the central nervous sytem. Life Sci 1981; 29:2345-55.
4. Fozard JR. Neuronal 5-HT receptors in the periphery. Neuropharmacology 1984; 23:1473-86.
5. Fozard JR, Mobarok Ali ATM. Blockade of neuronal tryptamine receptors by metoclopramide. Eur J Pharmacol 1989; 49:109-12.
6. Fozard JR, Gittos MW. Selective blockade of 5-hydroxytryptamine neuronal receptors by benzoic acid esters of tropine. Br J Pharmacol 1983; 80:511P.
7. Fozard JR. MDL 72222: a potent and highly selective antagonist at neuronal 5-hydroxytryptamine receptors. Naunyn-Schmiedeberg's Arch Pharmacol 1984; 326:36-44.
8. Fozard JR. The development and early clinical evaluation of selective 5-HT3 receptor antagonists. In: Fozard JR, ed. The Peripheral Actions of 5-Hydroxytryptamine. Oxford: Oxford University Press, 1989:354-76.
9. Richardson BP, Engel G, Donatsch P et al. Identification of serotonin M-receptor subtypes and their specific blockade by a new class of drugs. Nature 1985; 316:126-31.
10. Bradley PB, Engel G, Feniuk W et al. Proposals for the classification and the nomenclature of functional receptors for 5-hydroxytryptamine. Neuropharmacology 1986; 25:563-576.
11. Butler A, Hill JM, Ireland SJ et al. Pharmacological properties of GR38032F, a novel antagonist at 5-HT3 receptors. Br J Pharmacol 1988; 94:397-412.
12. Sanger GJ, Nelson DR. Selective and functional 5-hydroxytryptamine3 receptor antagonism by BRL 43694 (granisetron). Eur J Pharmacol 1989; 159:113-24.
13. Kilpatrick GJ, Jones BJ, Tyers MB. Identification and distribution of 5-HT3 receptors in rat brain using radioligand binding. Nature 1987; 330:746-48
14. Waeber C, Dixon K, Hoyer D et al. Localisation by autoradiography of neuronal 5-HT3 receptors in mouse CNS. Eur J Pharmacol 1988; 151:351-52.
15. Waeber C, Hoyer D, Palacios JM. 5-Hydroxytryptamine3 receptors in the human brain; autoradiographic visualization using [^{3}H]ICS 205-930. Neuroscience 1989; 31:393-400.
16. Barnes JM, Barnes NM, Costall B et al. Identification and characterization of 5-hydroxytryptamine3 recognition sites in human brain tissue. J Neurochem 1989; 53:1787-93
17. Costall B, Naylor RJ, Tyers MB. The psychopharmacology of 5-HT3 receptors. Pharmac Ther 1990; 47:181-202.

18. Jones BJ, Piper DC. 5-HT3 receptor antagonists in anxiety. In: King FD, Jones BJ, Sanger GJ eds. 5-Hydroxytryptamine3 Receptor Antagonists. Boca Raton: CRC Press, 1994:155-81.
19. Costall B, Naylor RJ. 5-HT3 receptor antagonism and schizophrenia. In: King FD, Jones BJ, Sanger GJ eds. 5-Hydroxytryptamine3 Receptor Antagonists. Boca Raton: CRC Press, 1994:183-201.
20. Costall B, Naylor RJ. 5-HT3 receptor antagonists in the treatment of cognitive disorders. In: King FD, Jones BJ, Sanger GJ eds. 5-Hydroxytryptamine3 Receptor Antagonists. Boca Raton: CRC Press, 1994:203-19.
21. Cooper SJ, Gilbert DB. 5-HT3 receptor antagonists, reward and drug withdrawal. In: King FD, Jones BJ, Sanger GJ eds. 5-Hydroxytryptamine3 Receptor Antagonists. Boca Raton: CRC Press, 1994:221-39.
22. Andrews PLR. 5-HT3 receptor antagonists and antiemesis. In:King FD, Jones BJ, Sanger GJ eds. 5-Hydroxytryptamine3 Receptor Antagonists. Boca Raton: CRC Press, 1994:255-317.
23. Lambert JJ, Peters JA, Hales TG et al. The properties of 5-HT3 receptors in clonal cell lines studied by patch-clamp techniques. Br J Pharmacol 1989; 97:27-40.
24. Derkach VA, Surprenant A, North RA. 5-HT3 receptors are membrane ion channels. Nature 1989; 339:706-9.
25. Peters JA, Malone HM, Lambert JJ. Recent advances in the electrophysiological characterization of 5-HT3 receptors. Trends Pharmacol Sci 1992; 13:391-97.
26. Peters JA, Lambert JJ, Malone HM. Electrophysiological studies of 5-HT3 receptors. In: King FD, Jones BJ, Sanger GJ eds. 5-Hydroxytryptamine3 Receptor Antagonists. Boca Raton: CRC Press, 1994:115-53
27. Boess FG, Martin IL. Molecular biology of 5-HT receptors. Neuropharmacology 1994; 33:275-317.
28. Lambert JJ, Peters JA, Hope AG. 5-HT3 receptors. In: North RA ed. Ligand and Voltage-Gated Ion Channels. Boca Raton: CRC Press, 1995:177-211.
29. Jackson MB, Yakel JL. The 5-HT3 receptor channel. Annu Rev Physiol 1995; 57:447-68.
30. Maricq AV, Peterson AS, Brake AJ et al. Primary structure and functional expression of the 5-HT3 receptor, a serotonin-gated ion channel. Science 1991; 254: 432-37.
31. McKernan RM. Biochemical properties of the 5-HT3 receptor. In: Hamon M ed. Central and Peripheral 5-HT3 receptors. London: Academic Press, 1992:89-102.
32. Hope AG, Downie DL, Sutherland L et al. Cloning and functional expression of an apparent splice variant of the murine 5-HT3 receptor A subunit. Eur J Pharmacol (Mol. Pharmacol. Sect.) 1993; 245: 187-92.
33. Werner P, Kawashima E, Reid J et al. Organization of the mouse 5-HT3 receptor gene and functional expression of two splice variants. Mol Brain Res 1994; 26: 233-41.
34. Boess FG, Lummis SCR, Martin IL. Molecular properties of 5-HT3 re-

ceptor-type binding sites purified from NG 108-15 cells. J Neurochem 1992; 59:1692-1701.
35. Hope AG. Cloning and functional expression of a murine 5-HT3 receptor subunit. PhD Thesis, The University of Dundee, U.K.
36. Johnson DS, Heinemann SF. Cloning and expression of the rat 5-HT3 receptor reveals species specific sensitivity to curare antagonism. Soc Neurosci Abs 1992; 18:249.
37. Isenberg KE, Ukhun IA, Holstad SG et al. Partial cDNA cloning and NGF regulation of a rat 5-HT3 receptor subunit. NeuroReport 1993; 5:121-24.
38. Belelli D, Balcarek JM, Hope AG et al. Cloning and functional expression of a human 5-hydroxytryptamine type $3A_S$ receptor subunit. Mol Pharmacol 1995; 48:1054-62.
39. Miyake A, Mochizuka S, Takemoto Y et al. Molecular cloning of a human 5-hydroxytryptamine3 receptor: heterogeneity in distribution and function among species. Mol Pharmacol 1995; 48:407-16.
40. Galzi J-L, Changeux J-P. Neurotransmitter-gated ion channels as unconventional allosteric proteins. Curr Op Structural Biol 1994; 4:554-65.
41. Karlin A, Akabas MH. Toward a structural basis for the function of nicotinic acetylcholine receptors and their cousins. Neuron 1995; 15:1231-44.
42. Green T, Stauffer K, Lummis SCR. Expression of recombinant homo-oligomeric 5-hydroxytryptamine3 receptors provides new insights into their maturation and structure. J Biol Chem 1995; 270:6056-61.
43. Boess FG, Beroukhim R, Martin IL. Ultrastructure of the 5-hydroxytryptamine3 receptor. J Neurochem 1995; 64:1401-5.
44. Morales M, Battenberg E, de Lecea L et al. Development and characterization of antibodies against the 5-HT3 receptor. Soc Neurosci Abs 1994; 20:474.3.
45. Kyte J, Doolittle RF. A simple method for displaying the hydropathic character of a protein. J Mol Biol 1982; 157:105-32.
46. Luthy R, Eisenberg R. Protein. In: Gribskov M, Devereux J eds. Sequence Analysis Primer. New York: Freeman, 1992:61-87.
47. Unwin N. Nicotinic acetylcholine receptor at 9Å resolution. J Mol Biol 1993; 1101-24.
48. Hucho F, Görne-Tschelnokow U, Strecker A. β-structure in the membrane-spanning part of the nicotinic acetylcholine receptor (or how helical are transmembrane helices?). Trends Biochem Sci 1994; 19:383-87.
49. Galzi J-L, Changeux J-P. Neuronal nicotinic receptors: molecular organisation and regulations. Neuropharmacology 1995; 34:563-82.
50. Yakel JL, Lagrutta A, Adelman JP et al. Single amino acid substitution affects desensitization of the 5-hydroxytryptamine type 3 receptor expressed in *Xenopus* oocytes. Proc Natl Acad Sci USA 1993; 90:5030-33.
51. Eiselé JS, Bertrand J-L, Galzi A et al. Chimaeric nicotinic-serotonergic receptor combines distinct ligand binding and channel specificities. Nature 1993; 366:479-83.
52. Pless DD, Lennarz WJ. Enzymatic conversion of proteins to glycoproteins. Proc Natl Acad Sci USA 1977; 74:134-38.

53. Miquel M-C, Emerit MB, Gingrich JA et al. Developmental changes in the differential expression of two serotonin 5-HT3 receptor splice variants in the rat. J Neurochem 1995; 65:475-83.
54. Uetz P, Abdelatty F, Villarroel A et al. Organisation of the murine 5-HT3 receptor gene and assignment to human chromosome 11. FEBS Lett 1994; 339:302-6.
55. Emerit MB, Martres MP, Miquel M-C et al. Differentiation alters the expression of the two splice variants of the serotonin 5-HT3 receptor-A mRNA in NG 108-15 cells. J Neurochem 1995; 65:1917-25.
56. Gordon JC, Rowland H. Nerve growth factor induces 5-HT3 recognition sites in rat phaeochromocytoma (PC12) cells. Life Sci 1990; 46:1435-42.
57. Furukawa K, Akaike N, Onodera H et al. Expression of 5-HT3 receptors in PC 12 cells treated with NGF and 8-Br-cAMP. J Neurophysiol 1992; 67:812-19.
58. Ortells MO, Lunt GG. Evolutionary history of the ligand-gated ion-channel superfamily of receptors. Trends Neurosci 18; 121-27.
59. Tecott LH, Maricq AV, Julius D. Nervous system distribution of the serotonin 5-HT3 receptor mRNA. Proc Natl Acad Sci USA 1993; 90:1430-34.
60. Johnson DS, Heinemann SF. Rat 5-HT3 receptor expression in the central and peripheral nervous systems - an in situ hybridization study. Soc Neurosci Abs 1993; 19:265.1
61. Waeber C, Pinkus LM, Palacios JM. The (S)-isomer of [^{3}H]zacopride labels 5-HT3 receptors with high affinity in rat brain. Eur J Pharmacol 1990; 181:283-87.
62. Barnes JM, Barnes NM, Champaneria S et al. Characterisation and autoradiographic localisation of 5-HT3 receptor recognition sites identified with [^{3}H]-(S)-zacopride in the forebrain of the rat. Neuropharmacology 1990; 29:1037-45.
63. Laporte AM, Kidd EJ, Vergé D et al. Autoradiographic mapping of central 5-HT3 receptors. In: Hamon M ed. Central and Peripheral 5-HT3 Receptors. London: Academic Press, 1992:157-87.
64. Wong EHF, Clark R, Leung E et al. The interaction of RS 25259-197, a potent and selective antagonist, with 5-HT3 receptors, in vitro. Br J Pharmacol 1995; 114:851-59.
65. Blandina P, Goldfarb J, Craddock-Royal et al. Release of endogenous dopamine by stimulation of 5-hydroxytryptamine3 receptors in rat striatum. J Pharmacol Exp Ther 1989; 251:803-9.
66. Battenburg E, Morales M, deLecea L et al. Immunocytochemical characterization of neurons expressing the 5-HT3 receptor. Soc Neurosci Abs 1994; 20:474.4.
67. Palacios JM, Waeber C, Mengod G et al. Molecular neuroanatomy of 5-HT receptors. In: Fozard JR, Saxena PR eds. Serotonin: Molecular Biology, Receptors and Functional Effects. Basel: Birkhäuser, 1991:5-20.
68. Kauer JA, McMahon LL. Hippocampal S. radiatum interneurons express functional serotonin receptors. Soc Neurosci Abs 1995; 21:242.3.

69. Ropert N, Guy N. Serotonin facilitates GABAergic transmission in the CA1 region of rat hippocampus in vitro. J Physiol 1991; 441:121-36.
70. Zeise ML, Batsche K, Wang RY. The 5-HT3 receptor agonist 2-methyl-5-HT reduces postsynaptic potentials in rat CA1 pyramidal neurons of the hippocampus in vitro. Brain Res 1994; 651:337-41.
71. Kawa K. Distribution and functional properties of 5-HT3 receptors in the rat hippocampal dentate gyrus: a patch-clamp study. J Neurophysiol 1994; 71:1935-47.
72. Piguet P, Galvan M. Transient and long-lasting actions of 5-HT on rat dentate gyrus neurones in vitro. J Physiol 1994; 481.3:629-39.
73. Leslie RA, Reynolds, DJM, Andrews, PLR et al. Evidence for pre-synaptic 5-hydroxytryptamine3 recognition sites on vagal afferent terminals in the brainstem of the ferret. Neuroscience 1990; 38:667-73.
74. Pratt GD, Bowery NG. The 5-HT3 receptor ligand [^{3}H] BRL 43694, binds to presynaptic sites in the nucleus tractus solitarius of the rat. Neuropharmacology 1989; 28:1367-76.
75. Larkman PM, Kelly JS. Pharmacological characterization of the receptor mediating 5-HT evoked motoneuronal depolarization in vitro. In: Fozard JR, Saxena PR eds. Serotonin: Molecular Biology, Receptors and Functional Effects. Basel: Birkhäuser, 1991:310-21.
76. Wilcox GL, Alhaider AA. Role of 5-HT3 receptors in pain modulation in the spinal cord. In: Hamon M ed. Central and Peripheral 5-HT3 Receptors. London: Academic Press, 1992:275-300.
77. Johnson DS, Heinemann SF. Detection of 5-HT3R-A, a 5-HT3 receptor subunit, in submucosal and myenteric ganglia of rat small interstine using in situ hydbridization. Neurosci Lett 1995; 184:67-70.
78. Champaneria S, Costall B, Naylor RJ et al. Identification and distribution of 5-HT3 recognition sites in the rat gastrointestinal tract. Br J Pharmacol 1992; 106:693-6.
79. Tecott L, Shtrom S, Julius D. Expression of serotonin-gated ion channel in embryonic neural and nonneural tissues. Mol Cell Neurosci 1995; 6:43-55.
80. Johnson DS, Heinemann SF. Embryonic expression of the rat 5-HT3 receptor subunit, 5-HT3R-A, in the rat: an in situ hybridization study. Mol Cell Neurosci 1995; 6:122-38.
81. Hussy N, Lukas W, Jones KA. Functional properties of a cloned 5- hydroxytryptamine ionotropic receptor subunit: comparison with native mouse receptors. J Physiol 1994; 418.2:311-23.
82. Gill CH, Peters JA, Lambert JJ. An electrophysiological investigation of the properties of a murine recombinant 5-HT3 receptor stably expressed in HEK 293 cells. Br J Pharmacol 1995; 114:1211-21.
83. Downie DL, Hope AG, Lambert JJ et al. Pharmacological characterization of the apparent splice variants of the murine 5-HT3 R-A subunit expressed in *Xenopus laevis* oocytes. Neuropharmacology 1994; 33:473-82.
84. Steward LJ, Boess FG, Steele JA et al. Agonist/antagonist recognition properties of the 5-HT3 receptor: importance of glutamate 106. Br J Pharmacol 1996; (In Press).

85. Hope AG, Brown AM, Peters JA et al. Characterization of a cloned human 5-HT3 receptor subunit stably expressed in HEK 293 cells. Br J Pharmacol 1995; 116:82P.
86. Yakel JL, Shao XM, Jackson MB. Activation and desensitization of the 5-HT3 receptor in a rat glioma x mouse neuroblastoma hybrid cell. J. Physiol 1991; 436:293-308.
87. Neijt HC, TeDuits IJ, Vijverberg, HPM. Pharmacological characterization of serotonin 5-HT3 receptor mediated electrical response in cultured mouse neuroblastoma cells. Neuropharmacology 1988; 27:301-7.
88. Boess FG, Sepulveda M-I, Lummis SCR et al. 5-HT3 receptors in NG 108-15 neuroblastoma x glioma cells: Effect of the novel agonist 1-(m-chlorophenyl)-biguanide. Neuropharmacology 1992; 31:561-64.
89. Fan P. Cannabinoid agonists inhibit the activation of 5-HT3 receptors in rat nodose ganglion neurons. J Neurophysiol 1995; 73:907-10.
90. Yang J. Ion permeation through 5-HT gated channels in N18 neuroblastoma cells. J Gen Physiol 1990; 96: 1177-98.
91. Yakel JL, Shao XM, Jackson MB. The selectivity of the channel coupled to the 5-HT3 receptor. Brain Res 1990; 533:46-52.
92. Yang J, Mathie A, Hille B. 5-HT3 receptor channels in dissociated rat superior cervical ganglion neurones. J Physiol 1992; 448:237-56.
93. Peters JA, Malone HM, Lambert JJ. An electrophysiological investigation of the properties of 5-HT3 receptors of rabbit nodose ganglion neurones in culture. Br J Pharmacol 1993; 110: 665-76.
94. Hargreaves AC, Lummis SCR, Taylor CW. Ca^{2+} permeability of cloned and native 5-hydroxytryptamine type 3 receptors. Mol Pharmacol 1995; 46:1120-27.
95. Gilon P, Yakel JL. Activation of 5-HT3 receptors expressed in *Xenopus* oocytes does not increase cytoplasmic Ca^{2+} levels. Receptors Channels 1995; 3:83-88.
96. Peters JA, Hales TG, Lambert JJ. Divalent cations modulate 5-HT3 receptor-induced currents in N1E-115 neuroblastoma cells. Eur J Pharmacol 1988; 151:491-95.
97. Lovinger DM. Inhibition of 5-HT3 receptor-mediated ion current by divalent metal cations in NCB-20 neuroblastoma cells. J Neurophysiol 1991; 66:1329-37.
98. Uki M, Narahashi T. Modulation of serotonin-induced currents by metals in mouse neuroblastoma cells. Soc. Neurosci Abs 1995; 21:679.6.
99. Belelli D, Hope AG, Peters JA et al. Functional properties of a human recombinant 5-HT3 receptor expressed in *Xenopus laevis* oocytes. Br J Pharmacol 1995; 116:229P.
100. van Hooft JA, Kooyman A, Verkerk A et al. Single 5-HT3 receptor-gated ion channel events resolved in N1E-115 mouse neuroblastoma cells. Biochem Biophys Res Commun 1994; 199:227-33.
101. van Hooft JA, Vijverberg HPM. Phosphorylation controls conductance of 5-HT3 receptor ligand-gated ion channels. Receptors Channels 1995; 3:7-12.

102. Robertson B, Bevan S. Properties of 5-hydroxytryptamine3 receptor-gated currents in adult rat dorsal root ganglion neurones. Br J Pharmacol 1991; 102:272-76.
103. Zhang L, Oz M, Weight FF. Potentiation of 5-HT3 receptor-mediated responses by protein kinase C activation. Neuroreport 1995; 6:1464-68.
104. Surprenant A, Matsumoto S, Gerzanich V. 5-HT3 receptors in guinea coeliac neurones. Soc Neurosci Abs 1991; 17:239.12.
105. Jones KA, Surprenant A. Single channel properties of the 5-HT3 subtype of serotonin receptor in primary cultures of rodent hippocampus. Neurosci Lett 1994; 174:133-36.
106. Shao XM, Yakel JL, Jackson MB. Differentiation of NG 108-15 cells alters channel conductance and desensitization kinetics of the 5-HT3 receptor. J Neurophysiol 1991; 65:630-38.
107. Sepúlveda M-I, Lummis SCR. 5-HT3 agonists can distinguish cloned 5-HT3 receptors from NCB20 and NG108-15 cell lines. Br J Pharmacol 1994; 112:316P.
108. Butler A, Elswood CJ, Burridge J et al. The pharmacological characterization of 5-HT3 receptors in three isolated preparations derived from guinea-pig tissues. Br J Pharmacol 1990; 101:591-98.
109. Newberry NR, Cheshire SH, Gilbert MJ. Evidence that the 5-HT3 receptors of the rat, mouse and guinea-pig superior cervical ganglion may be different. Br J Pharmacol, 1991; 102:615-20.
110. Peters JA, Malone HM, Lambert JJ. Characterization of 5-HT3 receptor mediated electrical responses in nodose ganglion neurones and clonal neuroblastoma cells maintained in culture. In: Fozard JR, Saxena PR, eds. Serotonin: Molecular Biology, Receptors and Functional Effects. Basel: Birkhauser, 1991:84-94.
111. Kilpatrick GJ, Barnes NM, Cheng CHK et al. The pharmacological characterization of 5-HT3 receptor binding sites in rabbit ileum: comparision with those in rat ileum and rat brain. Neurochem Int 1991; 19:389-96.
112. Wong EHF, Bonhaus DW, Wu I et al. Labelling of 5-hydroxytryptamine3 receptors with a novel 5-HT3 receptor ligand, [^{3}H]RS-42358-197. J Neurochem 1993; 60:921-30.
113. Pinkus LM, Gordon JC. Utilization of zacopride and its R- and S- enantiomers in studies of 5-HT3 receptor 'subtypes'. In: Fozard JR, Saxena PR, eds. Serotonin: Molecular Biology, Receptors and Functional Effects. Basel: Birkhauser, 1991:439-48.
114. Bonhaus DW, Wong EHF, Stefanich E et al. Pharmacolgical characterization of 5-hydroxytryptamine3 receptors in murine brain and Ileum using the novel radioligand [^{3}H]RS-42358-197: Evidence for receptor heterogeneity. J Neurochem 1993; 61:1927-32.
115. Wong EHF, Bonhaus DW, Lee JA et al. Different densities of 5-HT3 receptors are labeled by [^{3}H]quipazine, [^{3}H]GR 65630 and [^{3}H]granisetron. Neuropharmacology 1993; 32:869-75.
116. Perren MJ, Rogers H, Mason GS et al. A pharmacological comparison of

[^{3}H]-granisetron binding sites in brain and peripheral tissues of the mouse. Naunyn Schmiedeberg's Arch Pharmacol 1995; 351:221-28.
117. Peters JA, Malone HM, Lambert JJ. Antagonism of 5-HT3 receptor mediated currents in N1E-115 neuroblastoma cells by (+)-tubocurarine. Neurosci Lett 1990; 110:107-12.
118. Lovinger DM. Ethanol potentiation of 5-HT3 receptor-mediated ion current in NCB-20 neuroblastoma cells. Neurosci Lett 1991; 122:57-60.
119. Lovinger DM, White G. Ethanol potentiation of 5-hydroxytryptamine3 receptor-mediated ion current in neuroblastoma cells and isolated adult mammalian neurones. Mol Pharmacol 1991; 40:263-70.
120. Lovinger DM, Zhou Q. Trichloroethanol potentiation of 5-hydroxytryptamine3 receptor-mediated ion current in nodose ganglion neurones from adult rat. J Pharmacol Exp Ther 1993; 265:771-76.
121. Lovinger DM, Zhou Q. Alcohols potentiate ion current mediated by recombinant 5-HT3RA receptors expressed in a mammalian cell line. Neuropharmacology 1994; 33:1567-72.
122. Downie DL, Hope AG, Belelli D et al. The interaction of trichloroethanol with murine recombinant 5-HT3 receptors. Br J Pharmacol 1995; 114:1641-51.
123. Machu TK, Harris RA. Alcohols and anesthetics enhance the function of 5-hydroxytryptamine3 receptors expressed in *Xenopus laevis* oocytes. J Pharmacol Exp Ther 1994; 271:898-905.
124. Barann M, Göthert M, Fink K et al. Inhibition by anaesthetics of ^{14}C-guanidinium flux through the voltage-gated sodium channel and the cation channel of the 5-HT3 receptor of N1E-115 neuroblastoma cells. Naunyn-Schmiedeberg's Arch Pharmacol 1993; 347:125-32.
125. Miquel M-C, Emerit, MB, Gozlan H et al. Involvement of tryptophan residue(s) in the specific binding of agonists/antagonists to 5-HT3 receptors in NG108-15 clonal cells. Biochem Pharmacol 1991; 42:1453-61.
126. Schulte MK, Bloom KE, White MM. Evidence for the involvement of tryptophan in the binding of curare to 5-HT3 receptors. Soc Neurosci Abs 1995; 21:30.6.
127. Bertrand D, Devillers-Thiéry A, Revah F et al. Unconventional pharmacology of a neuronal nicotinic receptor mutated in the channel domain. Proc Natl Acad Sci USA 1992; 89:1261-65.
128. Filatov GN, White MM. The role of conserved leucines in the M2 domain of the acetylcholine receptor in channel gating. Mol Pharmacol 1995; 48:379-84.
129. Labarca C, Nowak MW, Zhang H et al. Channel gating governed symmetically by conserved leucine residues in the M" domain of nicotinic receptors. Nature 1995; 376:514-16.
130. Akabas MH, Kaufmann C, Archdeacon P et al. Identification of acetylcholine receptor channel-lining residues in the entire M2 segment of the α subunit. Neuron 1994; 13:919-27.
131. Bonhaus DW, Stefanich E Loury DN et al. Allosteric interactions

among agonists and antagonists at 5-hydroxytryptamine3 receptors. J Neurochem 1995; 65:104-10.
132. Hope AG, Jerman J, Jones BJ et al. Characterization by [^{3}H]granisetron binding of a cloned 5-HT3 receptor subunit (5-HT3R-A_S) stably expressed in HEK-293 cells. Can J Physiol Pharmacol 1994; 72:P18.4.25.
133. Akuzawa S, Miyata K, Fukutomi H. Characterization of [^{3}H]YM060, a potent and selective 5-HT3 receptor radioligand, in the cerebral cortex of rats. Eur J Pharmacol 1995; 281:37-42.
134. Hope AG, Peters JA, Brown AM et al. Characterization of a human 5-hydroxytryptamine3 receptor type A (h5-HT3R-A_S) subunit stably expressed in HEK293 cells. Br J Pharmacol 1996 (In Press).
135. Bufton KE, Steward LJ, Barber PC et al. Distribution and characterization of the [^{3}H]granisetron-labelled 5-HT3 receptor in the human forebrain. Neuropharmacology 1993; 32:1325-31.
136. Abi-Dargham A, Laruelle M, Wong DT et al. Pharmacological and regional characterization of [^{3}H]LY278584 binding sites in human brain. J Neurochem 1993; 60: 730-37.
137. Imoto K, Busch C, Sakmann B et al. Rings of negatively charged amino acids determine the acetylcholine receptor channel conductance. Nature 1988; 335:645-48.
138. Konno T, Busch C, Von Kitzing E et al. Rings of ionic amino acids as struictural determinants of ion selectivity in the acetylcholine receptor channel. Proc R Soc Lond 1991; B224:69-79.
139. Villaroel A, Herlitze S, Witzemann V et al. Asymmetry of the rat acetylcholine receptor subunits in the narrow region of the pore. Proc R Soc Lond 1992; B249:317-24.
140. Revah F, Bertrand D, Galzi J-L et al. Mutations in the channel domain alter desensitization of a neuronal nicotinic receptor. Nature 1991; 353:846-49.
141. Galzi J-L, Devillers Thiéry A, Hussy N et al. Mutations in the channel domain of a neuronal nicotinic receptor convert ion selectivity from cationic to anionic. Nature 1992; 359:500-505.

CHAPTER 8

STRUCTURE AND PHARMACOLOGY OF $GABA_A$ RECEPTORS

Ruth M. McKernan and Paul J. Whiting

8.1 INTRODUCTION

Gamma-aminobutyric acid (GABA) is the major inhibitory neurotransmitter in the vertebrate central nervous system, mediating synaptic inhibition through three receptor classes–the $GABA_A$, $GABA_B$ and $GABA_C$ receptors. $GABA_B$ receptors belong to the family of G-protein-coupled receptors which activate second messenger systems within the cell.[1] To date, these receptors have not been cloned.

$GABA_A$ and $GABA_C$ receptors belong to the family of ligand-gated ion channels. Binding of the agonist, GABA, to the receptor protein results in the rapid opening of the integral ion channel through which chloride enters the cell, hyperpolarizing the membrane, leading to synaptic inhibition. The two classes can be distinguished on the basis of their pharmacology: $GABA_C$ are insensitive to the $GABA_A$ antagonist bicuculine and the $GABA_B$ antagonist baclofen.[2] $GABA_C$ have been much less studied than $GABA_A$ receptors although information on their structure and function is now begining to emerge. They are currently proposed to be homo-oligomeric channels formed by ρ subunits of which there are two subtypes, ρ1 and ρ2.[3-5] These polypeptides are related to the genes which encode $GABA_A$ receptor subunits, and mRNA for their expression is predominantly located in the retina.[6]

This chapter will focus on the family of $GABA_A$ receptors, which form the largest GABA receptor class and have been much better characterized. The $GABA_A$ receptor is a pentameric protein with a plethora of modulatory binding sites whose activation can either potentiate or

Recombinant Cell Surface Receptors: Focal Point for Therapeutic Intervention, edited byMichael J. Browne.

inhibit the effects of GABA. Several of these sites have been particularly important since they are targets for some of the most widely used CNS-drugs. In this respect the $GABA_A$ receptor could be considered a pharmacological goldmine. The anaesthetic barbiturates such as pentobarbital, the benzodiazepines such as valium and temazepam, neurosteroid anaesthetics such as alphaxolone and the widely used drug, ethanol, all act through independent modulatory sites on the receptor to potentiate the effect of GABA. Further separate modulatory sites have also been described recently for other chemical structures, including the anticonvulsant, loreclezole, the antihelminthic compound ivermectin and the butyrolactones. Compounds which directly block the Cl^- channel, such as picrotoxin, have also been identified (for reviews see refs. 7-11).

The most studied site on the receptor is probably the benzodiazepine (BZ) binding site. Compounds acting through this site can either potentiate GABA (i.e. BZ agonists, which are as sedative, anticonvulsant and anxiolytic) or they can reduce the effect of GABA (i.e. BZ inverse agonists, which are proconvulsant). Compounds also exist which occupy the same benzodiazepine binding site but neither activate nor inhibit the effects of GABA (i.e. BZ antagonists). Pharmacological attempts to subdivide the large population of neuronal $GABA_A$ receptors further were made on the basis of the differential affinity of compounds for the better characterized benzodiazepine site rather than the agonist GABA site. Prior to cloning of the receptor it was accepted that there were at least three types of $GABA_A$ receptor in the CNS. Those with high affinity for compounds including zolpidem, CL-218,872 and some β-carbolines–the BZ1-subtype; those with lower affinity for these compounds, but with high affinity for flunitrazepam and Ro 15-1788–the BZ2 subtype[12] and those with low affinity for diazepam and most classical benzodiazepines–the diazepam-insensitive $GABA_A$ receptors.[13] Biochemical studies of the receptor supported the data obtained from pharmacological characterization of the benzodiazepine binding sites. Irreversibly labelling the receptor with photolabile compounds such as [^{3}H]flunitrazepam or [^{3}H]Ro 15-4513 produced several radioactive polypeptides which could be separated by gel electrophoresis. This was regarded as further evidence for receptor heterogeneity.[14]

Since the family of proteins which comprise the $GABA_A$ receptor have been cloned and assembled in recombinant systems, we can now appreciate the complexity of structures which were previously simplified into these three pharmacological classes. Consequently, the present classification of $GABA_A$ receptors into different subtypes is currently in a state of change. Therefore, for clarification in this chapter we will describe receptors by their subunit composition wherever possible.

8.2 CLONING AND EXPRESSION OF $GABA_A$ RECEPTORS

The first two subunits of the $GABA_A$ receptor were cloned by purifying large quantities of receptor from bovine brain to provide tryptic peptides which generated sufficient sequence to design oligonucleotide probes for screening of a bovine brain cDNA library. The two cDNA sequences identified were named α1 and β1.[15] It soon became clear that further cDNAs also existed; the first of these, with 79% identity to α1 and only 34% identity to β1 was thus referred to as α2. The cloning and sequencing of these cDNAs was the breakthrough in our understanding of the $GABA_A$ receptor. Subsequently, brain cDNA libraries were screened at low stringency with probes based on sequences conserved between these subunit cDNA and a whole family of $GABA_A$ receptor subunit cDNAs has now been identified (see Table 8.1). This work was primarily carried out by Seeburg and colleagues, but many other groups have also contributed to the task. The polypeptides are subdivided according to their relative amino acid identities. Thus within a subgroup (i.e. the α subunits) there is 70-80% identity, whereas between the subgroups (e.g. α and β subunits) there is only 30-40% identity. The regions of highest sequence identity are in the putative transmembrane domains and most of the putative N-terminal extracellular domain, while the regions of most diversity are the signal peptide and the large cytoplasmic loop region. To date there are six α-subunits, four β-subunits, three γ-subunits and one δ-subunit.

The amino acid sequences of the $GABA_A$ receptors show significant identity (20-30%) with other ligand gated ion channels such as the glycine receptor, the nicotinic acetylcholine receptor and the $5HT_3$ receptor (see chapters 6 and 8), and are thus part of the ligand-gated ion channel gene family (for review see ref. 17).

Further diversity of the 14 $GABA_A$ subunits is provided by alternative splicing. Within the $GABA_A$ receptor gene family alternative splicing has been so far been identified in 4 subunits: α6, β2, β4 and γ2. The splice variant of the α6 subunit leads to the omission of 10 amino acids from the N-terminal domain, and produces a non-functional receptor which has no obvious physiological role.[18] It appears to be present in about 20% of α6 mRNAs in the mouse cerebellum and arises apparently through incorrect splicing at the 3' splice site within the exon.

Two different alternatively spliced isoforms of the β2 subunit have been identfied in the chick brain[19] and the human brain.[20] In both cases the splice isoforms arise through the inclusion or omission of an additional exon in the putative cytoplasmic domain between TM3 and TM4. The exact position of the additional exon is different between the human and chicken species. In the chicken form of β2L the insert contains 17 amino acids, while in the human form the insert contains 38 amino acids. The functional or physiological consequence of the additional sequences are not known, but interestingly, in both cases they contain an additional putative phosphorylation site. The splice variant of β4 arises through an alternative 5' splice donor site, leading to the inclusion (β4') of an additional 4 amino acids in the putative large cytoplasmic domain.[21] Again any physiological role of the additional 4 amino acids has yet to be determined.

The γ2 variants also arise by inclusion (γ2L) or omission (γ2S) of an exon giving an additional 8 amino acids within the putative large cytoplasmic domain.[22,23] The γ2 splice variants are ubiquitous, being found in all species studied so far. Again the additional sequence contains a putative phosphorylation site for protein kinase C, suggesting a physiologically important role. Further, the γ2L splice variant has been suggested to be necessary for the potentiation of $GABA_A$ receptors by ethanol.[24,25]

8.3 CHROMOSOMAL ASSIGNMENT OF $GABA_A$ SUBUNIT GENES AND LINKAGE TO DISEASE STATES

Chromosome assignment for all the human $GABA_A$ receptor subunits is now known. Perhaps the most interesting observation is the clustering of genes. For instance, α1, β2 and γ2 are clustered,[26,27] and these subunits are known to coassemble to form an abundant receptor subtype.[28-31] Similarly, α6 and γ2 are clustered,[27,32] as are α2 and γ1.[33] These subunit pairs are also believed to coassemble with a β subunit to form major subunit populations.[34,35] This clearly suggests some sort of coordinated regulation of gene expression. However, this is obviously somewhat of an over-simplification since, for instance, the γ2 subunit is known to coassemble with other α subunits in addition to α1 and α6.[36-38]

The α5, β3 and γ3 genes are very closely linked on chromosome 15.[39,40] Indeed the γ3 and α5 genes are within 35 kb of each other.[41] These genes are also linked on chromosome 7 in the mouse.[39]

To date there has been no definitive linkage of a $GABA_A$ receptor subunit with a disease gene. Perhaps the closest linkage of a human disease is in the inherited disorder Angelman's syndrome[42] which has also been called 'laughing puppet's disease'. This syndrome is characterized by neurological dysfunctions (mental retardation, seizures, ataxia, absence of speech, jerky movements) and a smiling countenance. It has been mapped to the same locus (15q11,[43]) as the α5, β3 and γ3 genes.[44] Although the lesion is in the same region as the $GABA_A$ receptor subunit genes, whether a mutation in any of these genes is actually responsible for the disorder has yet to be established.

8.4 TRANSGENIC STUDIES

Animal models in which $GABA_A$ receptor subunits have been deleted or altered may prove to be very useful tools to understand the physiological role of the various $GABA_A$ receptor subtypes. As mentioned previously, the α5, β3 and γ3 genes are closely linked on chromosome 7 in the mouse. Mutant mouse lines have been generated which lack all three genes or just the α5 and/or γ3.[39,45] The mice lacking α5 and/or γ3 genes appear phenotypically normal, while mice lacking all three genes have a lethal cleft palate defect and those mice which do survive show a variable neurological phenotype including tremor and jerky gait. Detailed behavioral studies of these animals may provide information on the function of these subunits. One caveat to such experiments which must be borne in mind is that genes, in addition to those coding for α5, β3 and γ3, may also be deleted in these animals which could potentially confound interpretation of such work.

Generation of transgenic animals with altered $GABA_A$ receptor genes will undoubtedly prove to be a very useful tool. Mice have been generated in which the γ2 subunit has been specifically deleted.[46] The -/- mice have only 6% of the high affinity [^{3}H]flumazenil binding sites compared to wild-type animals (these are presumably receptors which contain a γ3 subunit) but 78% of the total $GABA_A$ receptors (as measured by [^{3}H]SR 95531 binding) which in electrophysiological experiments appeared to be αβ type receptors. The brains of these γ2 deleted animals appeared normal and had normal amounts and distribution of other receptor subunits. The phenotype was, however, lethal. Homozygous mice died shortly after birth, for reasons that were unclear, although the authors[46] speculated that lack of regulation by endogenous ligands of the BZ site may be to blame.

8.5 EXPRESSION OF RECOMBINANT $GABA_A$ RECEPTORS

Four different expression systems have been successfully used to produce recombinant $GABA_A$ receptors: *Xenopus* oocytes, baculovirus, transiently transfected mammalian cells and permanantly transfected mammalian cells.

The most widely used of these systems is certainly the *Xenopus* oocyte, which has become ubiquitious in the study of recombinant ion channels.[16,24,47,48] It has proved useful in the study of the functional properties and pharmacologies of different subunit combinations, and in structure-function analysis in combination with site-directed mutagenesis.[25,49-51] There has been no analysis of the structure of $GABA_A$ receptors expressed in oocytes (e.g. size of the assembled receptors, glycosylation of the subunits, etc.) but as far as can be determined they behave as would be predicted from previous studies on native receptors.

The baculovirus expression system has had much more limited utilization.[52,53] $GABA_A$ receptors produced by baculovirus infected cells appear to have the functional and pharmacological properties expected. Concerns over whether insect cells can appropriately synthesize and assemble these receptors thus appear unfounded, although confirmation that the receptors produced in baculovirus retain their structural integrity has yet to be provided. The baculovirus expression system has the potential to produce relatively high concentrations of membrane receptors, but for $GABA_A$ receptors this does not appear to be the case. A density of up to 3 pmol [^{3}H]flunitrazepam binding sites/mg protein has been reported,[52] which is of similar magnitude to expression levels produced in some transfected mammalian cell lines (up to 2.5 pmol [^{3}H]Ro 15-1788 binding sites/mg protein, Whiting and Hadingham, unpublished).

Transient expression of $GABA_A$ receptors in mammalian cell lines is a widely used technique.[54-56] It has the advantage over expression in *Xenopus* oocytes that receptors can be produced on a larger scale and thus can be used for ligand binding studies of recombinant receptors, as well as for functional studies. One could also speculate that since it is a mammalian system, unlike *Xenopus* oocytes or baculovirus, the cellular machinery will be more likely to synthesize and assemble receptors in a bona-fida manner. However, there is no evidence for this, and one could of course argue that a fibroblast cell used for transfection will not necessarily assemble receptors in the same way as a neuron.

Stable transfection of receptors on mammalian cells has not been a widely used approach for expression of $GABA_A$ receptors, or indeed ion-channels in general. This is presumably due to the time and effort required to make such cell lines. Stably transfected cells expressing $GABA_A$ receptors are of course not suitable for site directed mutagenesis experiments. They are, however, ideal for drug screening programs, and for longer term detailed pharmacological and functional studies and structural analysis of recombinant receptors. Evidence to date from such studies indicate that the $GABA_A$ receptors expressed by stably transfected fibroblast cells have structural, pharmacological and electrophysiological properties of native $GABA_A$ receptors.[57] The levels of stable expression of the recombinant receptors can vary according to the expression system,[58] but we have found that up to 2.5 pmols/mg protein can be achieved (Whiting and Hadingham, unpublished).

8.6 A COMPARISON OF NATIVE AND RECOMBINANT $GABA_A$ RECEPTORS

For monomeric receptors such as the G-protein receptor family, a comparison of recombinant and native receptors is relatively straightforward. However, for $GABA_A$ receptors the subunit composition of the many populations of receptors in the brain is largely unknown, therefore a detailed comparison of native and recombinant receptors with regard to their post-translational modification, phosphorylation, glycosylation, downregulation and desensitization has lagged behind that of other receptors.

It is known that recombinant $GABA_A$ receptors can be phosphorylated by cyclicAMP dependent protein kinase A (PKA), cyclicAMP dependent protein kinase C (PKC),[59] and tyrosine kinase.[60,61] Depending on the conditions used, modification by PKA can potentiate[62] or attenuate[63] GABA-gated currents. The attenuation is mediated by phosphorylation of Ser409 on the β1-subunit.[64] Phosphorylation by PKC can occur,[65] this being on residues Ser409 (on β1), Ser327 (on γ2S) and Ser343 (on γ2L)[64,65] modification of any one of which results in a decrease in the amplitude of the GABA-activated current.[65] $GABA_A$ receptors can be phosphorylated at tyrosine residues in the γ2-subunit resulting in increased whole cell currents.[60,61] Other kinases may also be able to phosphorylate the $GABA_A$ receptor; these may contribute to the reduced responses to GABA observed after repetitive stimulation in cell lines or oocytes, known as "run down"[67] which can be reduced by adding phosphorylation factors including ATP and magnesium.[68]

There are, in theory, greater than 100,000 possible subunit combinations that can be assembled to make a receptor. Some of the principles which appear to govern this process are beginning to be understood. It now seems well accepted that an intact receptor structure is composed of five subunits arranged around a central channel which conducts Cl^- and other monovalent anions. Three lines of evidence support the hypothesis that the $GABA_A$ receptor is pentamer–the similarity of sequence and structural motifs to the better characterized pentameric nicotinic receptor,[69] the physicochemical properties of the solubilized receptor[70] and the pentameric images of purified receptor preparations under the electron microscope.[71] As to the nature of which five subunits can combine to form a receptor, the picture is less clear. Early studies on the electrophysiological properties of recombinant receptors revealed that at least one α-subunit, one β-subunit and one γ-subunit are required to form a fully functional receptor in vivo.[72] If this represents a general blueprint for receptor assembly then there are still at least 10,000 possible subunit combinations available.

More recent studies have proposed that many receptors conform to a general plan being composed of two copies of an α-subunit, two copies of a γ-subunit and one β-subunit.[38,73,74] The α-subunits and the γ-subunits need not necessarily be of the same type (e.g. there are receptors which contain two different forms of α-subunit[36,74-78] and those which contain two different forms of γ-subunit[38,74]). To date there have been no reports in the literature of receptors which contain two different β-subunits, although in general, there have been fewer studies investigating the β-subunits, because their considerable sequence homology has limited the development of high titer β-selective antisera. This general blueprint for subunit assembly can accomodate the δ-subunit if it replaces a γ-subunit,[79] and would still allow small populations of receptors which contain both a δ and a γ-subunit.[37] This would reduce the number of potential combinations to about 200, although the upper limit on the number of different subunit combinations which actually exist if all the conflicting data in the literature is included has yet to be determined. Several recent reviews have been written which address this issue in more detail.[79,80] In this context it is important to note that an individual cell can express more than one subtype of receptor.[31,81,82] The mechanisms by which neurons assemble, sort and distribute receptor subtypes is currently unclear.

There is agreement on the structure of some of the major subtypes, e.g. $\alpha1\beta2\gamma2$, $\alpha2\beta1\gamma1$ and $\alpha3\beta2\gamma2$, which are particularly abundant in the CNS. For these and some of the other more minor subtypes the pharmacologies of the benzodiazepine site compare well with native receptor populations[54,76,83,84] (and see below).

There are many modulators of the $GABA_A$ receptor and much work has been carried out to determine which subunits, and in some cases which amino acid residues, contribute to their binding sites (see Table 8.1.[85]). Three lines of evidence show that the α-subunit contributes residues that are involved in forming the benzodiazepine binding site. Firstly, photolabile benzodiazepines such as [^{3}H]flunitrazepam and [^{3}H]Ro 15-4513 have been used to irreversibly label the benzodiazepine site of the $GABA_A$ receptor. Separation of the subunits by SDS-PAGE followed by immunoblotting identified the α-subunits as the primary target for photolabelling.[70,86,87] Secondly, recombinant receptors which differ only in their α-subunit composition have benzodiazepine sites with different pharmacologies. In combination with a β and γ2 subunit, receptors which contain an α1-subunit have the pharmacological profile previously described as BZ1.[72] Those containing an α2, α3 or α5 subunit have a BZ2 type pharmacology, although there are some compounds which can discriminate between the α2/3 subunits and α5,[83] and receptors which contain an α4 or α6 subunit in combination with a β and γ2 exhibit a pharmacology consistent with the previously described benzodiazepine-insensitive subtype.[13,31,34] Thirdly, site directed mutagenesis has since revealed at least some of the amino acid residues involved in forming the benzodiazepine binding site. His 100 in α1 (present in equivalent positions in α2, α3 and α5) is required for high affinity binding of benzodiazepine ligands[50] and gly 201 in the α1-subunit is required for high affinity binding of α1-selective ligands such as zolpidem and Cl-218,872.[88] A binding site for the GABA site has also been located on the α-subunit. Site directed mutagenesis[49] and photoaffinity labelling with the GABA site agonist [^{3}H]muscimol and subsequent microsequencing of radiolabelled peptide fragments[89] have both identified the residue as Phe-64.

The β-subunits also contribute to the GABA binding sites. The nature of the α or β-subunit results in variations in the degree of efficacy of a series of agonist and partial agonists.[90] Two domains have been identifed in the β2-subunit which are residues 157-160 (YGYT binding domain I) and residues 202-205 (TGSY, binding domain II).[91,92] At present it is not clear whether the α and β-subunits have separate and independent GABA binding sites or whether the binding site is formed at the interface between the α and β-subunits with contributions from both subunits. A further binding site for the anticonvulsant compound, loreclezole, has been identified on the β-subunit.[93] The GABA response of receptors which contain either a β2 or β3-subunit are sensitive to potentiation by loreclezole, whereas those which contain a β1-subunit are not. Site directed mutagenesis has established that the residue involved is Asn 289 (β2) which is located at the carboxy terminal end of the second transmembrane spanning domain.[94] Recent reports indicate that the β-subunit is also important for binding of the channel blocker, picrotoxin. Mutation of Thr 246 to Phe in the β2-subunit produces a picrotoxin-insensitive receptor.[95] By analogy with other ligand-gated ion channels this residue is predicted to be in the

Table 8.1. A summary of the properties of the GABA$_A$ receptor subunits

Subunit	amino acids/ splice variants	Phosphorylation sites	chromosome assignment	ligand binding sites
α1	456	PKC	5q32-33	BZ1 site-His100*, Gly201 (α1 selective compounds) GABA site-Phe64
α2	451	PKC	4p13-q11	BZ2 site, GABA site
α3	492	PKA,C	Xq28	BZ2 site, GABA site
α4	550	PKA,C	4p13-q11	Diazepam-insensitive BZ site, GABA site
α5	462	PKC	15q11-13	BZ2 site with low affinity for zolpidem, GABA site
α6	453	PKA,C	5q32-33	Diazepam insensitive BZ site, GABA site
β1	474	PKA,C	4p11-q13	GABA site
β2S/β2L	474/512	PKA,C	5q32-33	GABA site, loreclezole site-Asn 289
β3	473	PKA,C	15q11-q13	GABA site, loreclezole site-Asn 290
γ1	465	PKA,C,TyrK	4p13-q11	low affinity BZ site
γ2S/γ2L	468/476	PKA,C,TyrK	15q11-13	BZ site
γ3	467	PKA,C,TyrK	15q11-13	BZ site
δ	452	PKC	1p	BZ insensitive

Where two different sizes are given in the b2 and g2 subunits these arise by alternative splicing. Amino acids which contribute to the respective ligand binding sites are included where these are known. There are also likely to be additional residues which have not yet been determined. *His100 confers high affinity binding for benzodiazepines and the equivalent residue is also present in the a2, a3 and a5 subunits whereas an Arg is found at this position inthe a4 and a6 subunits.

second transmembrane spanning domain, which lines the pore of the Cl^- channel and supports the electrophysiological and biochemical data that picrotoxin directly blocks the ion channel of the receptor.

The γ-subunit is also involved in formation of the benzodiazepine binding site. One of the early observations with cloned $GABA_A$ subunits was that a γ2 subunit was necessary to recover functional responses to benzodiazepines and high affinity radioligand binding sites.[72] Both the γ2 and γ3 subunits are able to support high affinity benzodiazepine binding.[56,96] The γ1 subunit also forms a benzodiazepine binding site but generally, this has lower affinity for most benzodiazepine site ligands.[97,98] Residues in the γ2 subunit have been shown to determine the efficacy of benzodiazepine compounds.[99] Consistently, a portion of the [^{3}H]flunitrazepam photoincorporated into subunit polypeptides derived from recombinant α1β2γ2 receptors is immunoprecipitated by a γ2-antiserum.[87] No evidence has yet been found that the γ-subunit forms, or contributes to the formation of, a GABA binding site even as it has a role to play in mediating the effects of ethanol. For $GABA_A$ receptors to be potentiated by ethanol a γ2 subunit is required. More specifically the γ2L form is necessary, with its additional PKC phosphorylation site.[24] Whether this is actually the locus at which ethanol acts or whether the phosphorylation modifies the receptor in such a way as to make it sensitive to ethanol remains to be determined. At higher concentrations ethanol also interacts via another site on the receptor.[99]

To date the role of the δ-subunit has been less studied. It appears uninvolved in forming high affinity benzodiazepine binding sites and no other modulatory sites, have been detected which are unique to receptors containing this subtype.

There are several additional, and independent modulatory sites on the $GABA_A$ receptor whose subunit location has yet to be determined. These include binding sites for barbiturates, steroids and avermectins. The location of these sites has been harder to identify because they are present in most or all subunit combinations tested in model receptor systems in vitro. All doublet and triplet combinations so far tested have been potentiated by pentobarbital.[100,101] Although responses are observed with receptors containing each of the α subunits the degree of direct potentiation varies, with α6 producing the largest effect and α1 the smallest.[100] Barbiturates may also potentiate the receptor in the presence of GABA, through a separate binding site.[91] A similar lack of subunit specificity has been observed for the avermectin binding site which is also present in all subunit combinations thus far tested.[102] In radioligand binding assays using [^{3}H]Ivermectin the absolute affinities of a series of avermectin analogues varied depending on the type of α or β subunit present (McKernan et al, unpublished observation). A similar pattern emerges with steroid potentiation of the $GABA_A$ receptor (for review see ref. 103). 3α-OH-DHP and THDOC both potentiate receptors composed of homomeric β1, α1β1 and α1β1γ2.[104,105]

The type of α-subunit contributes to the degree of potentiation with a greater enhancement being observed with α3 than with α1, for example.[106] One interpretation of the relative lack of selectivity observed with barbiturates, steroids and avermectins is that the binding sites for these structures are formed by regions of the molecule which are identical or highly conserved within subunits. There have been no successful attempts to photoaffinity label the receptor with radioligands at any of these sites. One problem with such experiments is that steroids, barbiturates and avermectins are all very lipid soluble structures and, in general, radioligands at these sites have been technically difficult to work with for this reason. One could suggest that lipophilic regions of the receptor, potentially those including the transmembrane spanning domains, which are highly conserved, are involved in forming binding sites for these ligands.

8.7 FUTURE PROSPECTS

One important consequence of understanding which subunits combine to form intact receptors and the location of modulatory sites on those subunits is the ability to reconstitute individual subtypes in an artificial system and design subtype selective drugs. With such an abundance of modulatory sites on the receptor there is potential for generating subtype selective drugs which could have considerable advantages over the present generation of relatively non-selective compounds. Modulatory sites where the binding sites are known and there is clear evidence of specificity between subtypes, such as the benzodiazepine binding site and the loreclezole site, would provide the best targets. Where modulatory sites are located in highly conserved regions of the receptor it will be difficult to generate subtype selective drugs. The existence of benzodiazepines selective for the α1-containing (BZ1-subtype) was known before the receptor was cloned. However, since different isoforms of the α-subunit have been identified and expressed in reasonable amounts in cell lines, compounds have been identified which have some selectivity for receptors which contain an α5 subunit.[107] This would have been difficult prior to the revolution in the understanding of the molecular biology of GABA$_A$ receptors because the α5-containing receptors represent such a small fraction of all GABA$_A$ receptors in the brain.

With the development of cell systems expressing high levels of abundant GABA$_A$ receptor subtypes the path is now clear for medicinal chemists to synthesize subtype selective compounds. These will provide valuable tools to dissect the functions of the different GABA$_A$ subtypes in the brain. Compounds with selectivity for GABA subtypes might also be expected to produce GABAergic drugs, such as anxiolytics, sedatives and anticonvulsants with fewer side effects than the present generation of compounds. They may also provide drugs that act through GABA$_A$ receptor subtypes with completely novel therapeutic indications.

REFERENCES

1. Bowery NG. GABA-B receptor pharmacology. Ann Rev Pharmacol. Toxicol 1993; 33:109-147.
2. Quian H, Dowling JE. Novel GABA responses from rod-driven retinal horizontal cells. Nature 1993; 361:162-164.
3. Cutting GR, Curristin S, Zoghbi HY et al. Identification of a putative gamma-aminobutyric acid (GABA) receptor subunit p2 and colocalisation of the genes encoding p2 (GABARR1) and p1 (GABARR2) to human chromosome 6q14-q21 and mouse chromosome-4. Genomics 1992; 12:801-806.
4. Shimada S, Cutting G, Uhl GR. γ-aminobutyric acid A or C receptor? γ-aminobutyric acid p1 receptor RNA induces bicuculline, barbiturate and benzodiazepine insensitive γ-aminobutyric acid responses in *Xenopus* oocytes. Mol Pharmacol 1992; 41:683-687.
5. Kusama T, Spivak CE, Whiting P et al. Pharmacology of GABA p1 and GABA α1/β1 receptors expressed in *Xenopus* oocytes and COS cells Brit J Pharmacol 1993; 109:200-206.
6. Cutting GR, Lu L, O'Hara BF et al. (1991). Cloning of the gamma-aminobutyric acid rho1 cDNA: a GABA receptor subunit highly expressed in retina. Proc Natl Sci USA 1991; 88:2673-2677.
7. Sieghart W. Structure and pharmacology of γ-aminobutyric acid-A receptor subtypes. Pharmacol Revs 1995; 47 (2):181-234.
8. Stephenson FA. The $GABA_A$ receptors. Biochem J 1995; 310:1-9.
9. Whiting PJ, McKernan RM, Wafford KA. Structure and pharmacology of vertebrate $GABA_A$ receptor subtypes. Int Rev Neurobiol 1995; 38:95-137.
10. MacDonald RL and Olsen RW. $GABA_A$ receptor channels. Ann Rev Neurosci 1994; 17:569-602.
11. Olsen RW, Tobin AJ. Molecular biology of $GABA_A$ receptors. FASEB J 1990; 4:1469-1480.
12. Squires RF, Benson DI, Braestrup C et al. Some properties of brain specific benzodiazepine receptor: new evidence for multiple receptors. Eur J Pharmacol 1979; 10:825-830.
13. Sieghart W, Eichinger A, Richards JG et al. Photoaffinity labelling of benzodiazepine receptor proteins with partial inverse agonist [3H]Ro15-4513; a biochemical and autoradiographic study. J Neurochem 1987; 48:46-52.
14. Sieghart W, Drexler G. Irreversible binding of [^{3}H]flunitrazepam to different proteins in various brain regions. J Neurochem 1983; 41:47-55.
15. Schofield PR, Darlison MG, Fujita N et al. Sequence and functional expression of the $GABA_A$ receptor shows a ligand gated ion channel family. Nature 1987; 328:221-227.
16. Levitan ES, Schofield PR, Burt DA et al. Structural and functional basis fpr $GABA_A$ receptors heterogeneity. Nature 1989; 335:76-79.
17. Ortells MO, Lunt GG. Evolutionary history of the ligand-dated ion-channel superfamily of receptors. Trends in Neurosci 1995; 18:127-212.

18. Korpi ER, Kuner T, Kristo P et al. Small N-terminal deletion by splicing in cerebellar α6 subunit abolishes $GABA_A$ receptor function. J Neurochem 1994; 63:1167-1170.
19. Harvey RJ, Chinchetri MA et al. Alternative splicing of a 51 nucleotide exon that encodes a putative protein kinase C phosphorylation site generates 2 forms of the chicken γ-aminobutyric acidA receptor β2 subunit. J Neurochem 1994; 62:10-16.
20. McKinley DD, Lennon DJ, Carter DB. Cloning, sequence analysis and expression of two forms of mRNA coding for the human β2 subunit of the $GABA_A$ receptor. Mol Brain Res 1995; 28:175-179.
21. Bateson AN, Lasham A, Darlison MG. The chicken $GABA_A$ receptor α1 subunit: cDNA sequence and localisation of the corresponding mRNA. J Neurochem 1991; 56:1437-1440.
22. Whiting P, McKernan RM, Iversen LL. Another mechanism for creating diversity in γ-aminobutyrate type A receptors: RNA splicing directs expression of two forms of the γ2 subunit, one of which contains a protein kinase C phosphorylation site. Proc Natl Sci USA 1990; 87:9966-9970.
23. Kofuji K, Wang JB, Moss SJ et al. Generation of two forms of the γ-aminobutyric acid A receptor γ2-subunit in mice by alternative splicing. J Neurochem 1991; 56:713-715.
24. Wafford KA, Burnett DA, Leidenheimer NJ et al. Ethanol sensitivity of the $GABA_A$ receptor expressed in *Xenopus* oocytes requires eight amino acids contained in the γ2L subunit. Neuron 1991; 7:27-33.
25. Wafford KA, Whiting PJ. Ethanol potentiation of $GABA_A$ receptors requires phosphorylation of the alternatively spliced variant of the γ2 subunit. FEBS Lett 1992; 313:113-117.
26. Buckle VJ, Fujita N, Ryder-Cook AS et al. Chromosomal localisation of genes encoding the γ1 and γ2 subunits of the $GABA_A$ receptor indicates members of the gene family are often clustered in the genome. Neuron 1989; 3:647-654.
27. Wilcox AS, Warrington JA, Gardiner K et al. Human chromosomal localisation of genes encoding the γ1 and γ2 subunits of the $GABA_A$ receptor indicates members of the gene family are often clustered in the genome. Proc Natl Acad Sci 1992; 89:5857-5861.
28. Benke D, Mertens S, Trzeciak A et al. $GABA_A$ receptors display association of γ2 subunit with α1 and β2/3 subunits. J Biol Chem 1991; 266:4478-4483.
29. Benke D, Fritschy JM, Trzeciak A et al. Distribution, prevalence and drug binding profile of gamma-aminobutyric acid type A receptor subunits differing in their beta-subunit variant. J Biol Chem 1994; 269:27100-27107.
30. Laurie DJ, Seeburg PH, Wisden W. The distribution of 13 $GABA_A$ receptor subunit mRNAs in the rat Brain. II Olfactory bulb and cerebellum. J Neurosci 1992; 12:1063-1076.
31. Wisden W, Laurie DJ, Monyer HM et al. The distribution of 13 GABAA receptor subunit mRNAs in rat brain I. Telencephalon, diencephalon, mesencephalon. J Neurosci 1992; 12:1040-1062.

32. Hicks AA, Bailey MES, Riley BP. Further evidence for clustering of human $GABA_A$ receptor subunit genes: localisation of the α6-subunit gene (GABRA6) to distal chromosome 5q by linkage analysis. Genomics 1994; 20:285-288.
33. Wilcox AS, Berry R, Walter NAR et al. (personal communication).
34. Luddens H, Pritchett DB, Kohler M et al. Cerebellar $GABA_A$ receptor selective for a behavioral alcohol antagonist. Nature 1990; 346:648-651.
35. Quirk K, Gillard NP, Ragan CI et al. Model of subunit composition of γ-aminobutyric acid A receptor subtypes expressed in rat cerebellum with respect to their α and β/δ subunits J Biol Chem 1994a; 269:23 16020-16028.
36. Pollard S, Duggan MJ, Stephenson FA. Further evidence for the existence of a subunit heterogeneity within discrete γ-aminobutyric acid A receptor subpopulations. J Biol Chem 1993; 268:3753-3757.
37. Mertens S, Benke D, Mohler H. $GABA_A$ receptor populations with novel subunit combinations and drug binding profiles identified in brain by α5 and δ subunit specific immunopurification J Biol Chem 1993; 268:5965-5973.
38. Quirk K, Gillard NP, Ragan CI. γ-Aminobutyriuc acid type A receptors in rat brain can contain both γ2 and γ3 subunits, but γ1 does not exist in combination with another γ subunit. Mol Pharmacol 1994b; 45:1061-1070.
39. Nakatsu Y, Tyndale R, DeLorey TM et al. A cluster of 3 GABA receptor subunit genes is deleted in a neurological mutant of the mouse p-locus. Nature 1993; 364:448-450.
40. Wagstaff J, Chaillet JR, Lalande M. The GABAA receptor b3 subunit gene: characterisation of a human cDNA from chromosome 15q11q13 and mapping to a region of conserved synteny on mouse chromosome 7. Genomics 1991; 11:1071-1078.
41. Greger V, Knoll JHM, Woolff E et al. The gamma-aminobutyric acid receptor gamma-3 subunit gene is tightly linked to the alpha-5 subunit gene on human chromosome 15q11q13 and is transcribed in the same orientation. Genomics 1995; 26:258-264.
42. Angelman H. "Puppet children": a report on three cases. Devl Med Child Neurol 1965; 7:681-683l.
43. Williams CA, Gray BA, Hendrickson JE et al. Incidence of 15q deletion in the Angelman syndrome: a survey of twelve affected people. Am J Med Genet 1989; 32:339-345.
44. Sinnett D, Wagstaff J, Glatt K et al. High resolution mapping of the γ-aminobutyric acid receptor β3 and α5 gene cluster on chromosome 15q11-15q13, and localization of breakpoints in two Angelman Syndrome patients. Am J Hum Genet 1993; 52:1216-1229.
45. Culiat CT, Stubbs LJ, Montgomery CS et al. Phenotypic consequences of deletion of the γ3, α5 or β3 subunit of the type A γ-aminobutyric acid receptor of mice. Proc Natl Acad Sci USA 1994; 91:2815-2818.
46. Gunther U, Benson J, Benke D et al. Benzodiazepine insensitive mice generated by targeted disruption of the gamma2 subunit gene of

gamma-aminobutyric acid type A receptors. Proc Natl Acad Sci 1995; 17:7749-7753.

47. Puia G, Vicini S, Seeburg PH, Costa E. Influence of recombinant γ-aminobutyric acid A receptor subunit compositions on the action of allosteric modulators of γ-aminobutyric acid-gated chloride currents. Mol Pharmacol 1991; 39:691-696.
48. Woodward RM, Polenzani L, Miledi R. Characterization of bicuculline/baclofen-insensitive (ρ-like) γ-aminobutyric acid receptors expressed in *Xenopus* oocytes. II. Pharmacology of γ-aminobutyric acid A and γ-aminobutyric acid B receptor agonists and antagonists. Mol Pharmacol. 1993; 43:609-625.
49. Sigel E, Baur R, Kellenberger S, Malherbe P. Point mutations affecting antagonist and agonist dependent gating of $GABA_A$ receptor channels. EMBO 1992; 11(6):2017-2023.
50. Wieland HA, Luddens H, Seeburg PH. A single histidine in $GABA_A$ receptors is essential for benzodiazepine agonist binding. J Biol Chem 1992; 267:227-230.
51. Mihic SJ, Whiting PJ, Klein RL et al. A single amino acid of the human $GABA_A$ receptor γ2 subunit determines benzodiazepine efficacy. J Biol Chem 1994a; 269:32768-32773.
52. Carter DB, Thomsen DR, Im WB et al. Functional expression of $GABA_A$ chloride channels and benzodiazepine binding sites in baculovirus infected insect cells. Bio/Technology 1992; 10:679-681.
53. Prezenger JF, Im WB, Carter DB et al. Comparison of interactions of [^{3}H]muscimol, t-butylbicyclophosphoryo[^{35}S]thionate, and [^{3}H]flunitrazepam with cloned γ-aminobutyric acid A receptors of the α1β2 and α1β2γ2 subtypes. Mol Pharmacol 1993; 43:801-906.
54. Pritchett DB, Luddens H, Seeburg PH. Type I and type II $GABA_A$ benzodiazepine receptor produced in transfected cells. Science 1989; 245:1389-1392.
55. Smart TG, Moss SJ, Xie X et al. $GABA_A$ receptors are differentially sensitive to zinc-dependence on subunit composition. Brit J Pharmacol 1991; 103:1837-1839.
56. Luddens H, Seeburg PH, Korpi ER. Impact of β and γ variants on ligand binding properties of γ-aminobutyric acid type A receptors. Mol Pharmacol 1994; 45:810-814.
57. Hadingham KL, Harkness PC, McKernan RM et al. Stable expression of mammalian type A γ-aminobutyric acid receptors in mouse cells: demonstration of functional assembly of benzodiazepine-responsive sites. Proc Natl Acad Sci 1992; 89:6378-6382.
58. Hamilton BJ, Lennon D, Im HK et al. Stable expression of cloned rat $GABA_A$ receptor subunits in a human kidney cell line. Neurosci Lett 1993; 153:206-209.
59. Browning MD, Bureau M, Dudek EM et al. Protein kinase C and cAMP dependent protein kinase phosphorylate the β-subunit of the pruified $GABA_A$ receptor. Proc Natl Acad Sci USA 1990; 87:1315-1318.

60. Valenzuela CF, Machu TK, McKernan RM et al. Tyrosine kinase phosphorylation of $GABA_A$ receptors. Mol Brain Res 1995; 165-172.
61. Moss SJ, Gorrie GH, Amato A et al. Modulation of GABAA receptors by tyrosine phosphorylation. Nature 1995; 377:344-348.
62. Angelotti TP, Uhler MD, Macdonald RL. Enhancement of γ-aminobutyric acid type A receptor currents by chronic activation of cAMP-dependent protein kinase. Mol Pharmacol 1993; 44:1202-1210.
63. Moss SJ, Smart TG, Blackstone CD et al. Functional modulation of $GABA_A$ receptors by cAMP dependent protein phosphorylation. Identification of the cAMP-dependent protein kinase and protein kinase C phosphorylation sites within the major intracellular domains of the β1, γ2S and γ2L subunits of the γ-aminobutyric acid type A receptor. Science 1992a; 257:661-665.
64. Moss SJ, Doherty CA, Huganir RL. Identification of the cAMP dependent protein kinase and protein kinase phosphorylation sites within the major intracellular domains of the β1, γ2S and γ2L subunits of the γ-aminobutyric acid type A receptor. J Biol Chem 1992b; 267:14470-14476.
65. Kellenberger S, Malherbe P, Sigel E. Regulation of GABAA receptor function by protein kinase C phosphorylation. J Biol Chem 1992; 267:25660-25663
66. Krishek BJ, Xie X, Blackstone C et al. Regulatin of $GABA_A$ receptor function by protein kinase C phosphorylation. Neuron 1994; 12:1018-1095.
67. Stelzer A, Kay AR, Wong RKS. $GABA_A$ receptor function in hippocampal cells is maintained by phosphorylation factors. Science 1988; 241:339-341.
68. Gyenes M, Wang Q, Gibbs TT et al. Phosphorylation factors control neurotransmitter and neuromodulator actions at the g-aminobutyric acid type A receptor. Mol Pharmacol 1994; 46:542-549.
69. Unwin N. The structure of ion channels in the membranes of excitable cells. Neuron 1989; 3:565-575.
70. Stephenson FA, Duggan MJ. Understanding the $GABA_A$ receptor: a chemically gated ion channel. Biochem J 1988; 249:21-32.
71. Nayeem N, Green TP, Martin IL et al. Quaternary structure of the native $GABA_A$ receptor determined by electron microscopic image analysis. J Neurochem 1994; 62:815-818.
72. Pritchett DB, Sontheimer H, Shivers BH et al. Importance of a novel $GABA_A$ receptor subunit for benzodiazepine pharmacology. Nature 1989; 338:582-585.
73. Backus KH, Arigoni M, Drescher U et al. Stoichiometry of a recombinant $GABA_A$ receptor deduced from mutation-induced rectification. Neuroreport 1993; 5:285-288.
74. Duggan M, Pollard S, Stephenson FA. Immunoaffinity purification of $GABA_A$ receptor α-subunit iso-oligomers. Demonstration of receptor populations containing α1α2, α1α3 and α2α3 subunit pairs. J Biol Chem 1990; 266(36):24778-24784.
75. McKernan RM, Quirk K, Prince R et al. $GABA_A$ receptor subtypes

immunopurified from rat brain with α subunit specific antibodies have unique pharmacological properties. Neuron 1991; 7:667-676.
76. Luddens H, Killisch I, Seeburg PH. More than one α variant may exist in a $GABA_A$/benzodiazepine receptor complex. J Receptor Res 1991; 11:535-551.
77. Kern W, Sieghart W. Polyclonal anitbodies directed against an epitope specific for the α4-subunit of $GABA_A$ receptors identify a 67-kDa protein in rat brain membranes. J Neurochem 1994; 62:764-769.
78. Khan ZU, Guituerrez A, De Blas A. The subunit composition of a $GABA_A$/benzodiazepine receptor from rat cerebellum. J Neurochem 1994; 63:371-374.
79. Quirk K, Whiting PJ, Ragan CI et al. Characterization of δ-subunit containing $GABA_A$ receptors from rat brain. Eur J Pharmacol 1995; 290: 175-181.
80. McKernan RM, Whiting PJ. Which GABA subtypes really occur in the brain? TINS 1996; 19:139-143.
81. Baude A, Sequier J-M, McKernan RM et al. Differential subcellular distribution of the α6 subunit versus the α1 and β2/3 subunits of the $GABA_A$/benzodiazepine receptor complex in granule cells of the cerebellar cortex. Neurosci 1992; 51:739-748.
82. Fritschy JM, Benke D, Mertens S et al. Five subytpes of type A γ-aminobutyric acid receptors identified in neurons by double and triple immunofluorescence staining with subunit specific antibodies. Proc Natl Acad Sci 1992; 89:6726-6730.
83. Pritchett DB, Seeburg PH. Gamma aminobutyric $acid_A$ receptor α5 subunit creates novel type II benzodiazepine receptor pharmacology. J Neurochem 1990; 54:1802-1804.
84. Hadingham KL, Wingrove P, Le Bourdelles B et al. Cloning of cDNA sequences encoding human α2 and α3 γ-aminobutyric $acid_A$ receptor subunits and characterization of the benzodiazepine pharmacology of recombinant α1-, α2, α3 and α5-containing human γ-aminobutyric $acid_A$ receptors. Mol Pharmacol 1993a; 43:970-975.
85. Smith GB, Olsen RW. Functional domains of $GABA_A$ receptors. Trends in Pharm Sci 1995; 16:162-167.
86. Stephenson FA, Duggan MJ, Pollard S. The γ2 subunit is an integral component of the γ-aminobutyric $acid_A$ receptor but the α1 polypeptide is the principle site of the agonist benzodiazepine photoaffinity labelling reaction. J Biol Chem 1990; 265(53):21160-21165.
87. Stephenson FA, Duggan MJ. Mapping the benzodiazepine photoaffinity-labelling site with sequence-specific γ-aminobutyric $acid_A$ receptor antibodies. Biochem J 1989; 264:199-206.
88. Pritchett DB, Seeburg PH. γ-Aminobutyric acid type A receptor point mutation increases the affinity of compounds for the benzodiazepine site. Proc Natl Acad Sci 1991; 88:1421-1425.
89. Smith GB, Olsen RW. Identification of a $[^3H]$muscimol photoaffinity substrate in the bovine γ-aminobutyric $acid_A$ receptor α subunit. J Biol Chem 1994; 269:20380-20387.

90. Ebert B, Wafford KA, Whiting PJ et al. Molecular pharmacology of γ-aminobutyric acid type A receptor agonists and partial agonists in oocytes injected with different α,β, and γ receptor subunit combinations. Mol Pharmacol 1994; 46:957-963.
91. Amin J, Weiss DS. Molecular pharmacology of γ-aminobutyric acid type A receptor agonists and partial agonists in oocytes injected with different α,β, and γ receptor subunit combinations. Nature 1993; 366:565-569.
92. Amin J, Weiss DS. Homoeric p1 GABA channels: activation properties and domains. Recept Channels 1994; 2:227-236.
93. Wafford KA, Bain CJ, Quirk K et al. A novel allosteric modulatory site on the $GABA_A$ receptor β subunit 1994; Neuron 12:775-782.
94. Wingrove PB, Wafford KW, Bain CA et al. The modulatory action of loreclezole at the γ-aminobutyric acid type a receptor is determined by a single amino acid in the β2 and β3 subunits. Proc Natl Acad Sci 1994; 91:4569-4573.
95. Gurley DA, Amin J, Ross P et al. Point mutations in the M2 region of the $GABA_A$ receptor/ionophore complex abolish sensitivity to picrotoxin. Soc Neurosci Abstracts. 1994; 20:509.
96. Herb A, Wisden W, Luddens H et al. The modulatory action of loreclezole at the γ-aminobutyric acid type A receptor is determined by a single amino acid in the β2 and β3 subunits. Proc Natl Sci USA 1992; 89:1433-1437.
97. Ymer S, Draguhn A, Wisden W et al. $GABA_A$ receptor beta subunit heterogeneity: functional expression of clones cDNAs. EMBO J 1990; 9:3261-3267.
98. Wafford KA, Whiting PJ, Kemp JA. Differences in affinity and efficacy of benzodiazepine receptor ligands on recombinant $GABA_A$ receptor subtypes. Mol Pharmacol 1993; 43:240-244.
99. Mihic SJ, Whiting PJ, Harris RA. Anasthetic concentrations of alcohols potentiate $GABA_A$ receptor mediated currents: lack of subunit selectivity. Eur J Pharmacol (Mol Pharm) 1994b; 268:209-214.
100. Thompson SA, Whiting PJ, Wafford KA. Alpha subunits influence the action of pentobarbital on recombinant $GABA_A$ receptors. Brit J Pharmacol 1995; 117:521-527.
101. Hadingham KL, Wingrove PB, Wafford KA et al. The role of the beta subunit in determing the pharmacology of human $GABA_A$ receptors. Mol Pharmacol 1993b; 44:1211-1218.
102. Arena JP, Whiting PJ, Liu KK et al. Avermectins potentiate GABA-sensitive current in Xenopus oocytes expressing cloned $GABA_A$ receptors. Biophys J 1993; 64:A325.
103. Lambert JJ, Belelli D, Hill-Venning C et al. Neurosteroids and $GABA_A$ receptor function. Trends in Pharm Sci 1995; 16:295-303.
104. Puia G, Santi M, Vicini S et al. Neurosteroids act on recombinant $GABA_A$ receptors. Neuron 1990; 4:759-765.

105. Shingai R, Sutherland ML, Barnard EA. Effects of subunit types of the cloned $GABA_A$ receptor on the response to a neurosteroid. Eur J Pharmacol 1991; 206:77-80.
106. Lan NC, Gee KW, Bolger MB et al. Differential responses of expressed recombinant human γ-aminobutyric acid A receptors to neurosteroids. J Neurochem 1991; 57:1818-1821.
107. Gillard NP, Gu Zi-Qiang, Rice K et al. Substitution at position 8 of imidazo-benzodiazepine-6-ones is a determinant for $GABA_A$ α5-subtype selectivity. Brit J Pharmacol 1994.

CHAPTER 9

THE COMMON γ CHAIN AMONG RECEPTORS FOR THE CYTOKINES IL-2, IL-4, IL-7, IL-9 AND IL-15

Masataka Nakamura and Kazuo Sugamura

9.1 INTRODUCTION

In this chapter, we discuss the history, structure, function and pathogenesis of the common γ chain which is recruited into functional heteromeric receptor complexes for the cytokines interleukin 2 (IL-2), IL-4, IL-7, IL-9 and IL-15.

Growth and differentiation of hematopoietic lineage cells, which are central to a number of key biological processes involved in both normal and pathological processes, are under the control of direct cell-cell interaction through adhesion molecules and soluble chemical mediators, known as the cytokines. Failure of lymphocyte differentiation results in severe immunodeficiency. Patients with immunodeficiencies easily suffer serious infection and cancer. Treatment by either bone marrow transplantation or gene therapy is required for a radical cure of the patient. Disorder in cytokines and cell adhesion molecules is presumably associated with inflammation and autoimmune diseases. Any one cell lineage is affected by multiple cytokines, and most cytokines can influence cells of more than one lineage. Pleiotropy and redundancy are remarkable properties of cytokines, and may be explained by two mechanisms. The first mechanism is interaction between ligand and receptor: individual cells which belong to each lineage simultaneously express receptors for more than one cytokine. The second is

Recombinant Cell Surface Receptors: Focal Point for Therapeutic Intervention, edited by Michael J. Browne.

cross talk in signal transduction: different cytokine receptors may use the same or a similar set of cytoplasmic pathways for signaling.

The advent of expression cloning has revealed a new group of membrane glycoproteins, known as the cytokine receptor superfamily. Members of this family have common structural similarities in the extracellular region. A combination of multiple subunits generates functional receptor complexes with high affinities to cytokines. Of this family, three members (known as the common β chain, gp130 and common γ chain) have been demonstrated to be subunits which are shared among multiple heteromeric receptor complexes. Sharing of subunits may thus provide the molecular basis for the redundancy of cytokine action.

9.2 CYTOKINES AND CYTOKINE RECEPTORS

A large number of cytokines have been cloned. Production of recombinant cytokines in quantity has facilitated examination of the functions of cytokines in the immune system, hematopoietic system, nerve system and embryonic development. They exert their functions in pleiotropic manners through binding to specific receptors expressed on multiple target cells. A series of cytokine receptors have also been identified and molecularly cloned. Five subfamilies of cytokine receptor have been found so far based on the structural characteristics of their extracellular and intracellular domains: (i) the cytokine receptor superfamily; (ii) interferon (IFN) receptor family; (iii) tumor necrosis factor (TNF) receptor family; (iv) transforming growth factor β (TGF-β) receptor family; and (v) IL-8 receptor family.[1] The cytokine receptor superfamily is the largest family and contains at least 18 distinct receptor molecules of type I membrane glycoproteins with common features of (a) four conserved cysteine residues near the amino terminal and (b) the tryptophane-serine-x-tryptophane-serine (WSXWS) motif followed by the transmembrane segment (Fig. 9.1).[2,3] The other striking feature of members of the cytokine receptor superfamily is the absence of an intrinsic tyrosine kinase domain in the cytoplasmic region, which is common in receptors for growth hormones, such as platelet derived growth factor (PDGF), epidermal growth factor (EGF), insulin and fibroblast growth factor (FGF). Some members of the superfamily possess two characteristic intracytoplasmic sequences (box1 and box2) in the proximity of the membrane region.

Members of the superfamily associate as heteromers to form high-affinity receptors in a ligand-dependent fashion, which are able to transduce signals into the cytoplasm. The high-affinity receptors for cytokines are known to consist of a heterodimer with α and β subunits both of which belong to the cytokine receptor superfamily. On their own, generally, the α chains themselves have low affinity for ligand binding, and the β chains, unable to bind to ligands, act as signal transducers.

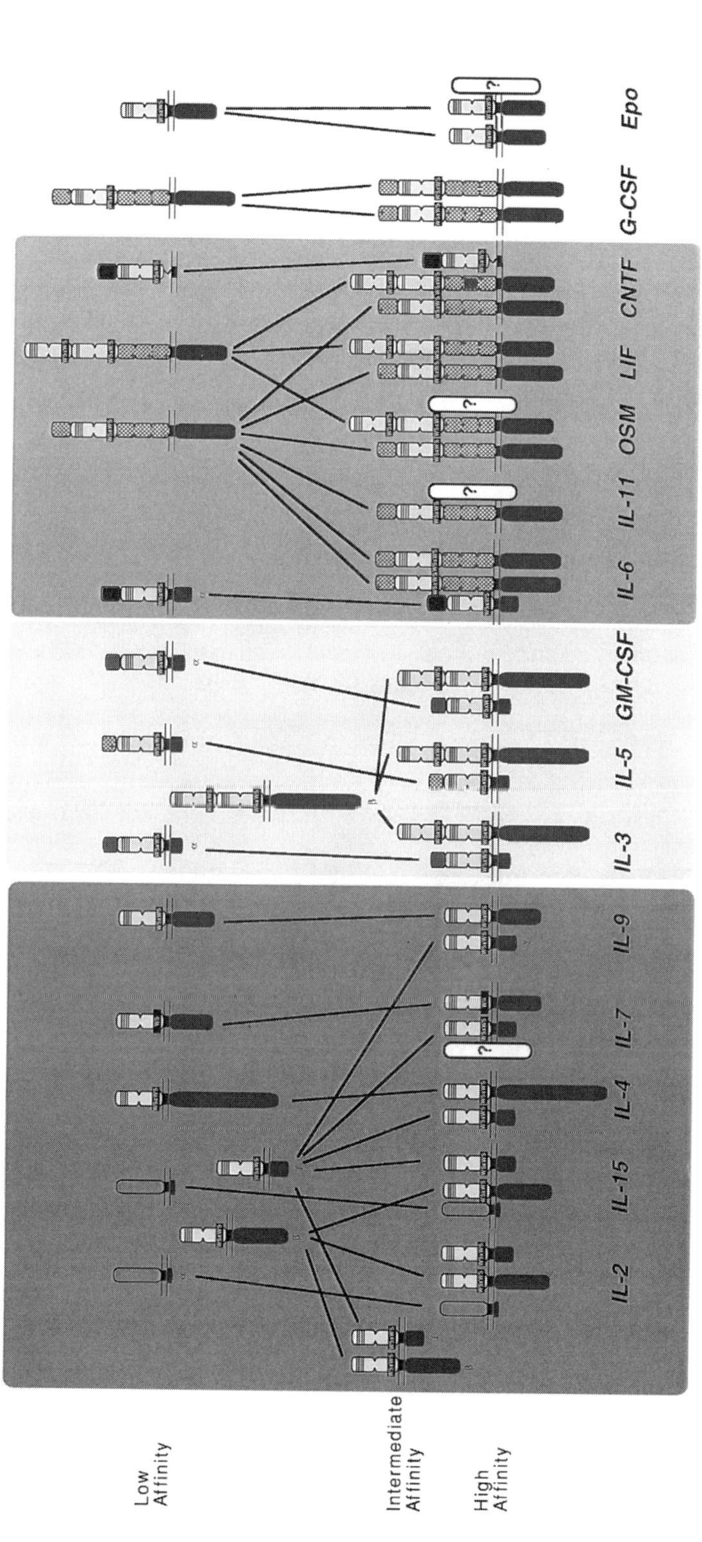

Fig. 9.1. Cytokine receptor superfamily.

The receptors for IL-6, IL-11, oncostatin M (OSM), ciliary neurotrophic factor (CNTF) and leukemia inhibitory factor (LIF) contain gp130. The common β chain is used in receptors for IL-3, IL-5 and granulocyte macrophage colony stimulation factor (GM-CSF). Both gp130 and common β chain essentially participate in increasing ligand binding affinities and in ligand-induced intracellular signal transduction, suggesting that the sharing of receptor subunits may in part reflect the redundant functions of cytokines.

9.3 IL-2/IL-2 RECEPTOR SYSTEM

9.3.1 T Cell Growth Factor

Activation of T cells is pivotal for determining immune reactions involving either T or B cells, or both cells. T cells initiate proliferation upon exposure to antigens. This process seems similar to that of fibroblast and epithelial cells stimulated with growth factors. Cytokines are simultaneously produced by activated T cells. They function on T cells themselves and on other types of immune cells. A series of experiments has revealed that activation of the T-cell receptor-CD3 complex makes T lymphocytes competent to proliferate. The growth factor induces proliferation of competent T cells.

IL-2 was initially found as a T cell growth activity in culture supernatant of T cells stimulated with antigens and mitogens.[4] Binding of IL-2 to a specific receptor on activated T cells results in proliferation of competent T cells. The IL-2 binding sites are seen not only on T cells but on other hematopoietic cells including B cells, natural killer (NK) cells and macrophages, and on glioma cells, indicating overlapping and pleiotropic effects of IL-2 as shown with other interleukins.

9.3.2 IL-2 Binding on T Cells

Most cells capable of IL-2 binding are demonstrated to have two distinct IL-2 binding sites with low (dissociation constants (Kd) $\sim 10^{-8}$ M) or high (Kd $\sim 10^{-11}$ M) affinities by Scatchard plot analysis.[5] T cells infected with human T-cell leukemia virus type-I (HTLV-I) constitutively express IL-2 receptor, like activated normal T cells. Immunization with human T cell lines infected with HTLV-I generated a monoclonal antibody recognizing the Tac antigen.[6] The monoclonal antibody partially inhibited IL-2 binding: it inhibited low- and high-affinity IL-2 binding, however, even in the presence of the monoclonal antibody, significant IL-2 binding still remains with Kd of $\sim 10^{-9}$ M, which is intermediate between the high and low affinity states. Little or no growth inhibition is seen in IL-2-dependent cultures following addition of the monoclonal antibody. Thus, three different affinity sites on IL-2 reactive cells were revealed. Importantly, IL-2 concentrations that bind to the IL-2 binding sites with intermediate and high affinities are the same as the concentrations of IL-2 that are required for IL-2-dependent growth.[8,9]

A complete cDNA clone encoding the Tac antigen was isolated.[10-12] The cDNA clone for the Tac antigen (p55) exhibits only low affinity (Kd ~10^{-8} M) to IL-2 binding, when expressed on various types of cells negative for IL-2 receptor: and the Tac antigen is now known as the IL-2 receptor α chain (IL-2Rα). The peptide backbone of IL-2Rα consists of 251 amino acid residues with no significant homology to other cytokine receptors. The cytoplasmic domain of IL-2Rα contains only 13 amino acid residues, which seems insufficient to harbor a signal transducing ability. In fact, the low-affinity receptor is nonfunctional in transduction of IL-2-mediated intracellular signals.

Further experiments in which cell surface molecule was cross-linked after IL-2 treatment, detected, rather than the α chain, at least two other molecules cross-linked with IL-2 and predicted that those may be components for the intermediate- and/or high-affinity receptor subunits.[13-17] We and others independently developed monoclonal antibodies which interfere with IL-2 binding to the intermediate- and high-affinity binding sites but not to the low-affinity binding site, and recognize a 75 kDa molecule (p75), presumably the largest one among the molecules cross-linked with IL-2.[18,19] Treatment of cells with an anti-p75 monoclonal antibody together with anti-IL-2Rα antibody results in marked inhibition of IL-2-dependent cell proliferation. The complete cDNA clone was isolated by expression cloning using the monoclonal antibody, and is known as the IL-2 receptor β chain (IL-2Rβ).[20]

The mature form of IL-2Rβ is deduced to have 525 amino acid residues with two pairs of the conserved cysteine residues and the WSXWS motif, belonging to the cytokine receptor superfamily (Fig. 9.2). The cytoplasmic domain with 286 amino acid residues is long enough to be presumed capable of intracellular signal transduction and contains unique regions: box1, a serine-rich region, an acidic region and a proline-rich region. IL-2Rβ participates, when expressed on lymphoid lineage cells, in formation of the intermediate-affinity receptor. Expression of IL-2Rβ along with IL-2Rα on lymphoid cells confers the same high affinity as that on activated T cells and HTLV-I infected T cells.[20]

In general, high-affinity receptors for cytokines are known to be complexes containing α and β subunits, both of which belong to the cytokine receptor superfamily. The IL-2 receptor complex is, however, different from receptors for other interleukins in that IL-2Rα, having only the 13 amino acids in its cytoplasmic region, is not a member of the cytokine receptor superfamily. Furthermore, besides the low- and high-affinity receptors, the intermediate-affinity IL-2 receptor does appear to be expressed under physiological conditions.

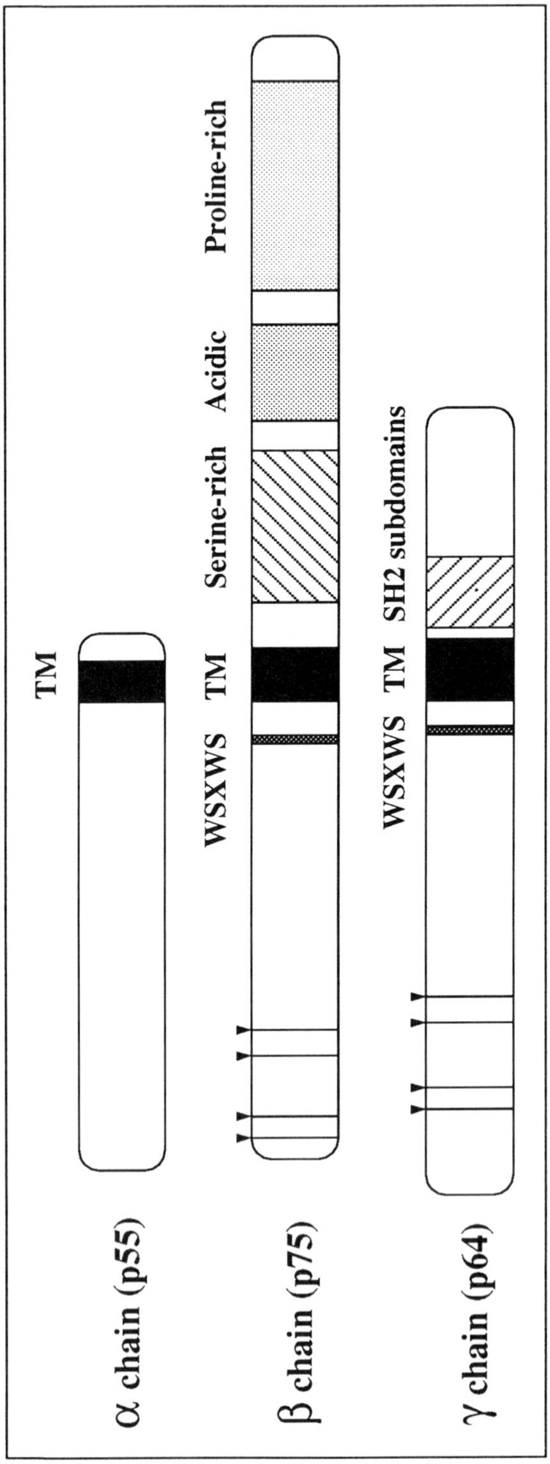

Fig. 9.2. Schematic representation of the IL-2 receptor subunits. TM indicates transmembrane region. The left and right regions of TM represent extracellular and intracellular domains, respectively. Solid triangle and WSXWS indicate the conserved cysteine residues and WS motif commonly seen in the cytokine receptor superfamily.

9.3.3 The Third Component

Interestingly, results obtained with fibroblast cells transfected with the IL-2Rβ cDNA clone are different from those with transfected lymphoid cells.[21] IL-2Rβ expressed alone on fibroblast cells does not show any appreciable binding to IL-2. By contrast, when IL-2Rβ is expressed along with IL-2Rα, transfected fibroblast cells bind to IL-2 with Kd of ~10^{-10} M, which is between intermediate and high affinities. These results indicate that IL-2Rα and IL-2Rβ chains are not sufficient to generate the intermediate- and high-affinity receptors capable of signal transduction, and suggests that the IL-2 receptor contains a component that is specifically and endogenously expressed in lymphoid cells.

In parallel, we demonstrated that an anti-IL-2Rβ antibody, TU11, which does not interfere with IL-2 binding, precipitates IL-2Rβ in lysates of cells not treated with IL-2, while it precipitates a 64 kDa cell surface molecule (p64) distinct from IL-2Rα, along with IL-2Rβ from the cell lysates of lymphoid cells treated with IL-2.[22] Furthermore, the amount of p64 coprecipitated is proportional to the number of intermediate-affinity receptors. The associated molecule (p64) is not seen in the immunoprecipitates from even IL-2Rα and IL-2Rβ transfectants of non-lymphoid cell lines. These findings suggest the possibility that the p64 molecule may be the lymphoid-specific third component of the IL-2 receptor, which contributes to formation of the high- and intermediate-affinity IL-2 receptors. The complete cDNA clone encoding the p64 molecule was isolated with a probe generated by polymerase chain reaction with primers designed from the amino acid sequences. The encoded product is known as the γ chain of the IL-2 receptor.[23]

9.4 THE γ CHAIN

9.4.1 Structural Properties

Analysis of nucleotide sequences of the γ chain cDNA revealed that the mature form consists of 347 amino acid residues with the sequence typical of the cytokine receptor superfamily (Fig. 9.2). The extracellular domain contains a leucine zipper-like region, which may be capable of promoting association of subunits. The function of the remainder of the extracellular domain is unknown. The cytoplasmic domain consisting of 86 amino acid residues contains a region homologous to the last two subdomains (out of a total of 5) of the Src homology region 2 (SH2).[24] The SH2 domain is believed to mediate intracellular signals involving tyrosine kinases through binding phosphotyrosine residues.[25] The last two subdomains of the SH2 domain present in the γ chain seem to be insufficient for binding to phosphotyrosine residues, however, it appears to be important for intracellular signal transduction (see below).

cDNA clones encoding the mouse γ chain were isolated based on cross-hybridization with the human γ chain cDNA probe.[26,27] The predicted structure of the mouse γ chain is identical to that of human γ chain except for length in both the extracellular and cytoplasmic regions; the mouse γ chain extracellular region is one amino acid longer, while the cytoplasmic region is one amino acid shorter than the human γ chain. The homology between the two species at the amino acid level is 66% in the extracellular domain and 83% in the cytoplasmic domain, including the SH2-like region.

9.4.2 Reconstitution of the IL-2 Receptor

Insight into function of the γ chain in IL-2 binding was gained using mouse fibroblast L cells transfected with the γ chain along with IL-2Rα and/or IL-2Rβ.[23] Expression of IL-2Rα alone or both IL-2Rα and γ chain elicited low affinity (Kd ~ 10^{-8} M) binding to IL-2. Either IL-2Rβ or γ chain alone possesses undetectable affinities (Kd > 10^{-7} M). The association rate constant with the αβγ heterotrimer complex is four-fold larger than that with the αβ heterodimer complex, and the dissociation rate constant is one-fifth of that with the αβ complex, thus generating two different high-affinity receptors with a Kd of 10^{-11} M and 10^{-10} M, respectively.[28] The latter αβ complex is nonfunctional for intracellular signal transduction and is referred to as the pseudo-high affinity IL-2 receptor.[21,23] These results illustrate that the γ chain is involved in mechanisms by which IL-2 associates with and dissociates from receptors, resulting in the generation of the high-affinity IL-2 receptor (Kd ~ 10^{-11} M) together with IL-2Rα and IL-2Rβ, and the intermediate-affinity IL-2 receptor (Kd ~ 10^{-9} M) together with IL-2Rβ. The dissociation rate constants of the αβγ and βγ complexes on lymphoid cells are generally much slower than those on fibroblastoid cells,[28] suggesting that there is a significant difference between these two types of cells in regulating the dissociation of IL-2 from the receptors.

Immunoprecipitation experiments with TU11, anti-IL-2Rβ antibody, confirm association of the γ chain with IL-2Rβ in an IL-2-dependent fashion. It should be noted that the association is seen with the L transfectants expressing either IL-2Rβ and γ chain or IL-2Rα, IL-2Rβ and γ chain, but not with the L transfectants expressing either IL-2Rα and IL-2Rβ or IL-2Rα and γ chain, although they are able to bind to IL-2. The result demonstrates that IL-2Rα, although it contributes to increase in the affinity to IL-2, is not critically involved in physical association between IL-2Rβ and the γ chain.

Internalization of IL-2 has been shown to be associated with the functional IL-2 receptors. Only the L transfectants with IL-2Rβ and γ chain together internalize IL-2, independent of IL-2Rα.[23]

9.4.3 Difference in βγ Heterodimer Between Human and Mouse

Reconstitution studies of mouse IL-2 receptors provide evidence of some differences in the IL-2 receptor system between human and mouse, although the homologies of the IL-2 receptor subunits between mouse and human at the amino acid level are 61%, 58% and 70% for IL-2Rα, IL-2Rβ and γ chain, respectively.[26,27,29-31] A critical difference is found between the two species in the IL-2 binding affinity of the βγ heterodimer complex. The mouse αβγ heterotrimer complex expressed on either lymphoid or fibroblastoid transfectants comprises the high-affinity receptor for both human and mouse IL-2, while, in contrast to the human system, the mouse βγ heterodimer complex exhibits undetectable affinity for IL-2 binding.[26] Mouse splenic $CD8^+$ T cells, expressing significant amounts of IL-2Rβ and γ chain but not IL-2Rα, have no ability to bind IL-2.[32] Therefore, exogenous expression of the IL-2Rα and IL-2Rβ chains together into a mouse T cell line, BW5147, which expresses the γ chain endogenously induces the ability to bind IL-2 with Kd of 90 nM. However, introduction of IL-2Rβ solely does not show any appreciable binding. On the other hand, not only transfectant cells with both human IL-2Rβ and γ chain but also human $CD8^+$ T cells and NK cells, which express both IL-2Rβ and γ chain but not IL-2Rα, are able to be activated for their cytotoxicity and proliferation in response to IL-2.[33-35] A mouse T lymphoma cell line, EL-4, expresses both IL-2Rβ and γ chain, but no IL-2 binding is seen. Introduction of exogenous human IL2-Rβ expression makes the cells capable of IL-2 binding. Moreover, $CD8^+$ T cells from mouse which carries a transgene of the human IL-2Rβ are able to response to IL-2, demonstrating a critical difference between mouse and human IL-2Rβ chains. These results suggest that IL-2Rα is necessary for formation of the functional IL-2 receptor in the mouse system; the functional IL-2 receptors are consisting of the αβγ or βγ complexes in human, and of the αβγ complex in mouse.

9.4.4 Expression of the γ Chain

Two-color flow cytometry study demonstrates that, among human peripheral blood cells, the γ chain is seen on all of the cell populations including $CD4^+$ T, $CD8^+$ T, $CD20^+$ B, and $CD56^+$ NK cells and $CD14^+$ monocytes (Fig. 9.3).[36] Most of the granulocytes also express the γ chain. IL-2Rα and IL-2Rβ are differentially expressed on these cell populations, and their expression is enhanced by extracellular stimuli such as antigens and mitogens. $CD8^+$ T and $CD56^+$ NK cells significantly express IL-2Rβ but little of IL-2Rα, while $CD4^+$ T cells little or faintly express IL-2Rβ. The γ chain is also expressed in all human cell lines tested from hematopoietic lineage except for THP-1, a myelomonocytic cell line.[37]

Profiles of IL-2Rα, IL-2Rβ and γ chain expressions on various populations of mouse splenic cells are substantially the same as those on the human populations.[38] Mouse thymus includes the double negative ($CD4^-CD8^-$), double positive ($CD4^+CD8^+$) and single positive ($CD4^+$ or $CD8^+$) T cell subsets. The double negative T cell subset—the most

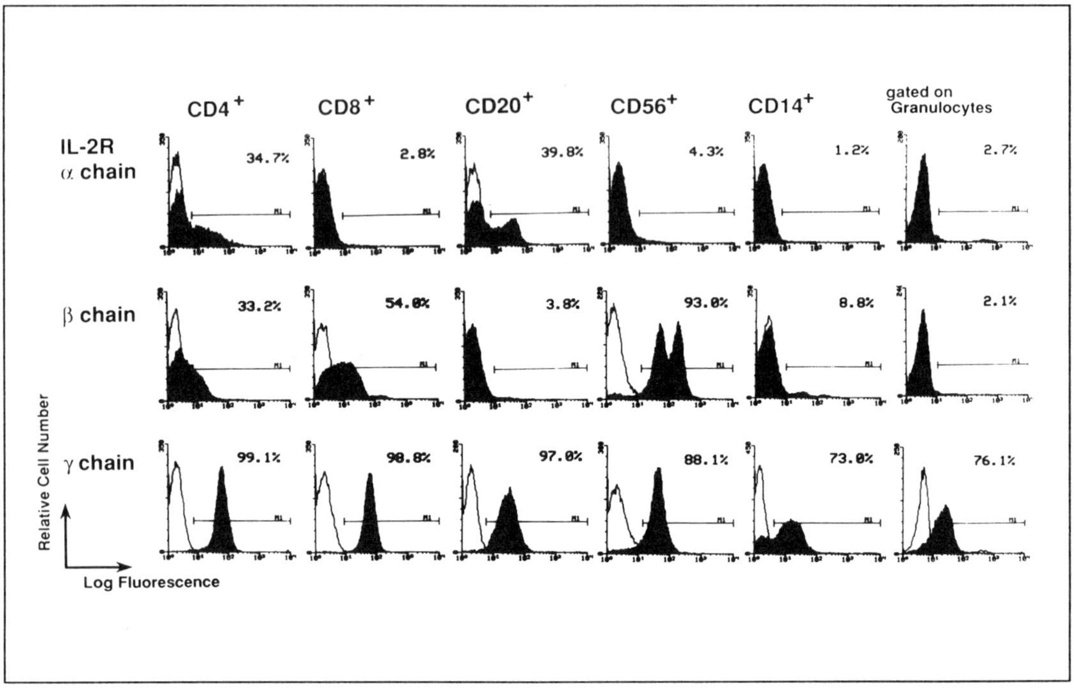

Fig. 9.3. Expression of the IL-2 receptor subunits on hematopoietic lineage cells. (Courtesy of Int. Immunol. 1994; 6:1273-7) Human peripheral blood leukocytes were stained with lineage marker (CD4, CD8, CD20, CD56 and CD14) antibodies and antibodies for the IL-2 receptor subunits, and analyzed with a flow cytometer. CD4⁺ and CD8⁺ are T lymphocytes. CD20⁺ is B lymphocytes. CD56⁺ and CD14⁺ are NK cells and monocytes, respectively. Reprinted from Int Immunol 1995; 6:1273, by permission of Oxford University Press.

immature subset—displays a high level of expression of the γ chain, IL-2Rα and IL-7Rα, but little or no expression of IL-2Rβ and IL-4Rα, predicting that the double negative T cells express the functional IL-7 receptor. The double positive subset is positive for IL-4Rα, but negative for IL-2Rα, IL-2Rβ and IL-7Rα, therefore they could have no functional receptors for IL-2 and IL-7. CD8⁺ T cells express IL-2Rβ, γ chain, IL-4Rα and IL-7Rα, suggesting expression of functional receptors for IL-2, IL-4 and IL-7. On the other hand, CD4⁺ T cells also express the γ chain, IL-4Rα and IL-7Rα, but little or no IL-2Rα and IL-2Rβ, rendering expression of functional receptors for IL-4 and IL-7.

9.4.5 Regulation of γ Chain Expression

The γ chain is generally detected on the hematopoietic lineage cells but scarcely on non-hematopoietic cells. To elucidate the regulatory mechanisms of the γ chain expression, the promoter region of the γ chain has been isolated and examined for its transcriptional activity.[37,39] The 633-base pair fragment upstream of the initiation codon for the γ chain shows the promoter activity, which contains four possible P.U.1 recognition sites, one PEA-3 site, three GT boxes, one TATA-like sequence and one Ets-binding site. 12-O-Tetradecanoylphorbol 13-acetate (TPA) or phytohemagglutinin (PHA) but not forskolin significantly increases transcription from the γ chain gene promoter. In contrast, IL-2 treatment induces suppression of transcription from the γ chain gene promoter. IL-2-induced suppression of IL-2R expression may be one of mechanisms by which activation of T cells by antigen recognition is transient and returns to background level immediately after an immune reaction. Such a suppressive effect of IL-2 could not be observed in HTLV-I-infected T cells. A *trans*-acting transcriptional activator, HTLV-I Tax, is believed to be involved in leukemogenesis, at least in part, through aberrant expression by *trans*-activation of cellular genes related to cell growth,[40] which include the IL-2Rα gene.[41] Tax is proven to augment expression of the γ chain, suggesting that Tax nullifies the IL-2-mediated suppression of γ chain expression by the *trans*-activation.[36]

In contrast to T cells, a low level of γ chain expression on fresh human monocytes is enhanced by either IL-2 or IFNγ stimulation, presumably due to the stabilization of γ chain mRNA.[42]

9.5 XSCID AND γ CHAIN MUTATION

The genomic structure of human γ chain exhibits eight exons and seven introns spanning 4.2 kb which is mapped on chromosome Xq12-13.[43] Surprisingly, the study indicates a causative relation between the γ chain and a severe immune disease, X-linked severe combined immunodeficiency (XSCID). The location of the γ chain gene is the locus for XSCID. Furthermore, mutations are detected in the γ chain gene of three patients with XSCID.[43] The same results were independently observed by several laboratories, confirming that all the patients with XSCID carry mutant genes for the γ chain.[44-47] The γ chain mutations were thus demonstrated to be causatively related to XSCID.

Severe combined immunodeficiency (SCID) features profound depression of both humoral and cell-mediated immunity, these are classified into several forms according to their genetic bases. Among SCID, linkage analyses revealed that the loci of putative genes on X chromosome are responsible for XSCID, Wiskott-Aldrich syndrome (WAS), X-linked lymphoproliferative syndrome (XLP), X-linked hyper IgM syndrome (XHM) and X-linked chronic granulomatous disease (CGD).[48] As many as 50% of patients with SCID appear to have XSCID, which manifests complete or profound T cell and NK cell deficit, but carry normal or slightly increased numbers of B cells. This leads to death mostly within a year after birth if not treated with bone marrow transplantation.[49,50] The γ chain is therefore necessary for development of T and NK cells.

Mutations found in the γ chain gene of XSCID vary from patient to patient, and spread over the γ chain, including regions for the WSXWS motif, conserved cysteine residues and SH2-like domain. The mutations are in the extracellular and cytoplasmic domains caused by base substitutions leading to missense and nonsense mutations, frame shift mutations by one and two base deletions and deletions of exons. The γ chain mutants found in patients with XSCID have been shown to be unable to mediate signals for growth in response to IL-2, even though some of them retain the ability to bind IL-2.[45,46,51]

B cells in patients with XSCID appear normal in terms of number, indicating that B cell development, at least by the IgM-positive B cell stage, seems independent of the γ chain function. However, a very low amount of immunoglobulin is present in serum of patients with XSCID. One may presume that B cells would be incapable of secreting immunoglobulin, because of lack of T cell help in these patients. Alternatively, mutations in the γ chain may affect B cell function. The cooperation in vitro between B cells from patients with XSCID and T cells from normal individuals is able to lead the B cells to immunoglobulin production.[52] On the other hand, obligatory carrier females who are normal in immunological properties show non-random inactivation of the X chromosome in mature B cells contrary to random activation in normal individuals.[53] This is also true in T and NK cells in carrier females.[54,55] Immature B cells expressing mutant γ chains may suffer disadvantage at conversion to mature B cells. These observations may reflect crucial functions of the γ chain in B cells as well as T cells.

Two atypical XSCID phenotypes carrying normal or reduced numbers of T cells also have been reported to have mutations in the γ chain gene. In the first case, a normal number of $CD4^+$ and $CD8^+$ T cells in the periphery was reported,[56] but the peripheral T cells were poorly functional. Two different mRNA are transcribed from the γ chain gene; one is nonfunctional due to aberrant RNA splicing, and the other functional in spite of a mutation, with a molecular ratio of approximately 4:1, suggesting a small amount of the γ chain is sufficient for T cell development but not enough for T cell activation. The second case exhibits less severe immunodeficiency with a reduced number of T cells in both $CD4^+$ and $CD8^+$ populations and normal numbers of NK and B cells,[57] however, the T cells do not show the capacity to proliferate in response to PHA and there was an impaired production of immunoglobulins to a specific antigen, diphtheria toxin. An intriguing observation is that normal numbers of T and B cells are present in the periphery of a dog (basset hound) carrying the nonfunctional γ chain[58,59] but the T cells fail to respond to PHA. Overall, these results imply functional implication of the γ chain in T cell development could be different from that in T cell activation in immune responses.

The question then is whether the impairment of T cell development in XSCID is primarily caused by the dysfunction of the IL-2 receptor. The question is answered in part by the observation that IL-2-deficient SCID patients generate a normal number of T cells in the periphery.[60-63] In addition, IL-2 deficient mice generated by gene targeting similarly show normal T cell development.[64,65] An indication from these observations is that the dysfunction of the γ chain in the IL-2 receptor is not directly associated with a defect of T cell development in XSCID, further suggesting the possibility that the γ chain is a component of functional receptor complexes for cytokines other than IL-2. This suggestion is reminiscent of the common β chain and gp130. This hypothesis is consistent with the fact that a wider range of cells express the γ chain than those affected by IL-2.

9.6 SHARING OF THE γ CHAIN IN MULTIPLE RECEPTORS

The assumption that the γ chain is a common subunit of receptor complexes for IL-2 and other cytokines that are necessary for early T cell development was tested by us and others with different approaches. We took advantage of monoclonal antibodies specific for the γ chain. The others used a way of reconstitution with cDNA for receptor subunits. Most cytokine receptors are known to be complexes. At that time when we cloned the γ chain, though binding subunits of the receptor for IL-4, IL-7, IL-9 and erythropoietin (Epo) had been described no associated subunits had been identified. We speculated that the binding subunits associate with the γ chain to form fully functional receptor complexes.

We first developed two types of monoclonal antibodies, TUGm2 and TUGm3, specific for the mouse γ chain; TUGm2 can block specific interaction between IL-2 and the γ chain, and TUGm3 can precipitate the mouse γ chain cross-linked with IL-2.[66]

Lymphoid cells expressing IL-4Rα have been shown to exhibit higher affinities than those of non-lymphoid COS-7 transfectant with the IL-4Rα gene,[67,68] suggesting that the high-affinity IL-4 receptor is a complex composed of at least IL-4Rα and another subunit, a possible molecule, the γ chain. To examine this possibility, CTLL-2 cells, for which growth is dependent on either IL-2 or IL-4, were used. Treatment of CTLL-2 cells with TUGm2 reduces the affinity of IL-4 binding from the high affinity (Kd=130 pM) to the intermediate affinity (Kd=850 pM), and the affinity to IL-2 binding from the high (Kd=130 pM) to the pseudo-high (Kd=370 pM).[66] No significant change in number of IL-4 binding sites is seen. The γ chain itself binds little or no IL-4. As expected, TUGm2 markedly suppress IL-4-dependent growth

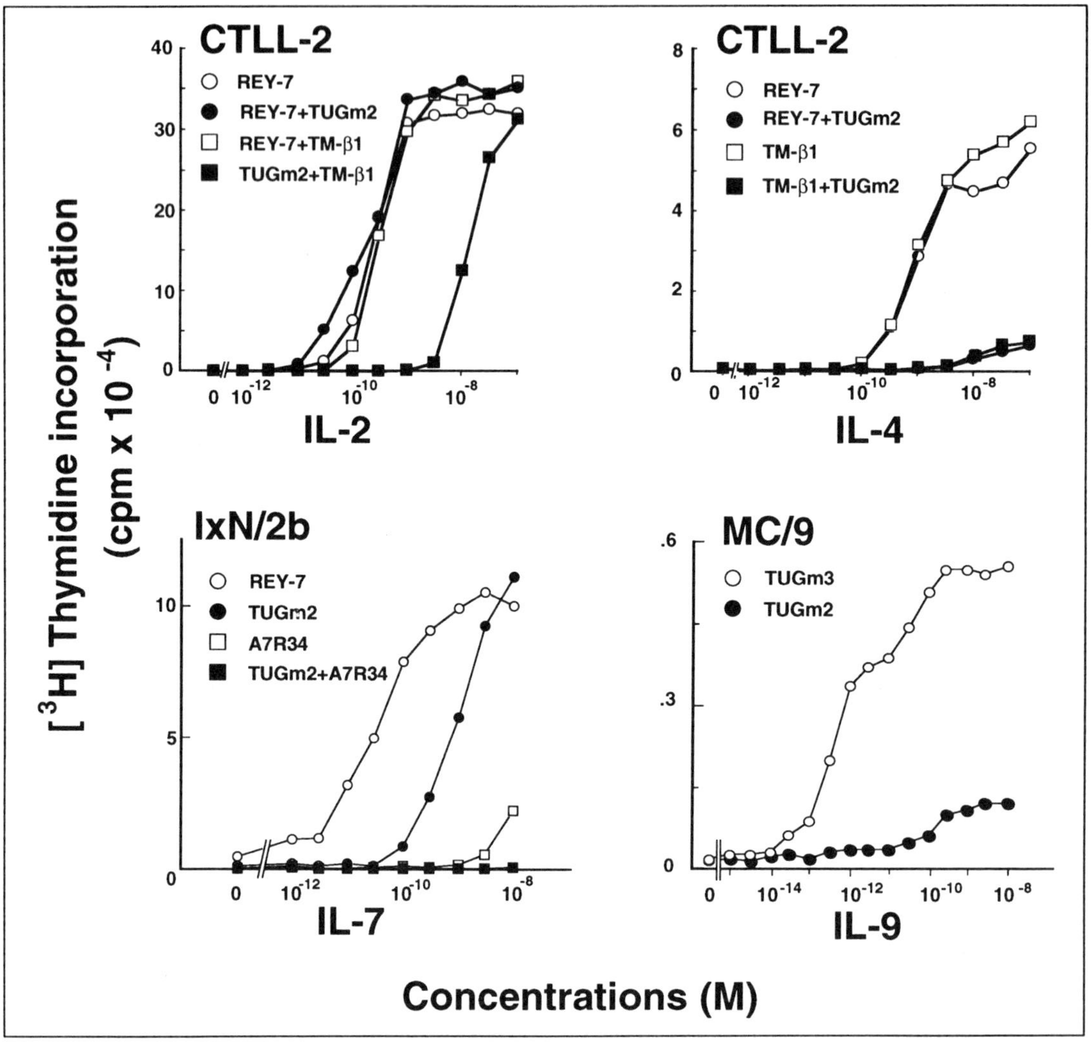

Fig. 9.4. Inhibitory effects of anti-γchain antibody on cell growth induced by IL-2, IL-4, IL-7 and IL-9. (Courtesy of Science 1993; 262:1874-7 and 1994; 263:1453-4, and Int. Immunol. 1995; 7:115-20) Cells were cultured in the medium containing interleukins in the presence of antibodies. Cell growth was monitored by [^{3}H] thymidine incorporation. CTLL-2 is a mouse IL-2-dependent T cell line. IxN/2b and MC/9 are mouse IL-7 responsive pre-B cell line and IL-3-dependent mast cell line, respectively. TUGm2, TM-β1 are anti-mouse γ chain and IL-2 receptor β chain monoclonal antibodies, respectively. A7R34 is an anti-mouse IL-7 receptor α chain monoclonal antibody. TUGm3 is an anti-mouse γchain monoclonal antibody which does not interfere with IL-2 binding. REY-7 is an anti-human parvovirus VP-1 monoclonal antibody. Reprinted with permission from Science 1993; 262:187.

of CTLL-2 cells (Fig. 9.4). These observations indicate that the γ chain is necessary for IL-4-mediated cell growth. Immunoprecipitation with TUGm3 of the γ chain after chemical cross-linking with IL-4 confirms direct participation of the γ chain in the IL-4 receptor complex. These results suggest a physical association of the γ chain with the functional IL-4 receptor. The same conclusion is reached by experiments of the reconstitution of the IL-4 receptor by co-transfection of the γ chain and IL-4Rα genes.[69]

IL-4 was initially detected as a B cell growth factor, and later shown to possess a broader spectrum of biological activities, which include growth stimulation of the HT-2 T cell line and the MC/9 mast cell line, and of resting T cells by costimulation with TPA.[70,71]

There are reports suggesting involvement of IL-4 in T cell development; however, IL-4-deficient mice show normal T and B cell development though with profoundly reduced serum levels of IgG1 and IgE.[72] Parenthetically, sharing of a subunit between receptors for IL-4 and IL-13 has been suggested based on the observations that IL-13 competitively inhibits binding of IL-4 to the IL-4 receptor on a cell line responsive to both IL-4 and IL-13.[73,74] However, no evidence that the γ chain is shared with the IL-13 receptor complex has been obtained.

Similar to the IL-4Rα chain, transfection of the IL-7Rα gene into COS-7 cells induces IL-7 binding with Kd of 270 pM, which is lower than that of the IL-7 receptor expressed on lymphoid cells. The γ chain was thus suspected to be shared with the IL-7 high-affinity receptor. The Scatchard plot analysis revealed that TUGm2 reduces the affinity of IL-7 binding to IxN/2b cells, a mouse IL-7 responsive pre-B cell line, from 79 pM to 255 pM without affecting the number of the binding sites, whereas no change in the low affinity (Kd=16 nM) is seen.[75] It should note that the reduced affinity with TUGm2 is distinct from the low affinity state, presumably suggesting involvement of another molecule distinct from the IL-7Rα and γ chain subunits. In this context, it is interesting to learn that cells blocked with A7R34, anti-IL-7Rα antibody, alone or together with TUGm2 display only the low-affinity receptor. Growth of concanavalin A (ConA)-stimulated splenic cells and IxN/2b cells is significantly blocked by addition of TUGm2 (Fig. 9.4). Simultaneous addition of TUGm2 and A7R34, completely blocked the IL-7-dependent cell growth. Thus, IL-7-dependent growth requires function of the γ chain, demonstrating functional participation of the γ chain in the functional IL-7 receptor complex. The chemical cross-linking experiments show the direct interaction between the γ chain and IL-7. Reconstitution experiments with the γ chain and IL-7Rα genes show the same result.[76]

IL-7 derived from stromal cells stimulates the proliferation of pre-B cells ($B220^+$, sIg^-) and induces growth of double negative thymocytes, suggesting function of IL-7 on T-cell development. Administration of a neutralizing monoclonal antibody specific for IL-7 into mice results in substantial reduction of all major thymic subpopulations.[77,78] IL-7Rα chain deficient mice provide critical information that development of T and B cells is markedly inhibited in their early developmental stages, resulting in a severe immunodeficiency. Therefore one may assume that signals from IL-7 are responsible for T cell development. The question of whether early T cell development requires IL-7 was addressed using the mouse thymic organ culture system. The most immature thymocytes, pro-T cells, with phenotypes of Pgp1$^+$, c-kit$^+$, and CD4$^-$ CD8$^-$ were sorted from day-15 fetal thymus and cultured with fetal thymic lobes (eliminated for preexisting lymphocytes by treatment with deoxyguanosine). Ten days later, single positive cells, CD4$^+$CD8$^-$ or CD4$^-$CD8$^+$, in addition to double positive CD4$^+$CD8$^+$ cells, are generated. Transition from the double-negative CD4$^-$CD8$^-$, Pgp1$^+$ c-kit$^+$ pro-T cells into the double positive CD4$^+$CD8$^+$ T cells was completely inhibited in the culture containing anti-γ chain monoclonal antibody (TUGm2) and anti-IL-7Rα (A7R34).[80] This result indicates a functional role for IL-7 in T cell development and suggests a direct connection between XSCID and lack of IL-7 function due to dysfunction of the γ chain. However, a dissimilarity in phenotypes between human XSCID and IL-7 dysfunction mice is seen: a normal number of B cells in patients with XSCID vs. significantly reduced number of T and B cells in mice treated with anti-IL-7 antibody.[81,82] Moreover, a similar reduction in T and B cells was observed in mice carrying a deficiency of either IL-7 gene or IL-7Rα gene.[79,83]

IL-9 is reported to have the mitogenic activity for T cells, B cells and fetal thymocytes but not for adult thymocytes.[85,86] The functional IL-9 receptor complex was also shown to contain the γ chain by a reconstitution study.[51] Independently, we showed that a mouse IL-3-dependent mast cell line, MC/9, which is also responsive to IL-9, is completely inhibited for IL-9-induced growth by treatment with TUGm2 (Fig. 9.4).[84] Similar inhibition of IL-9-mediated cell growth by TUGm2 is seen with a mouse IL-2-dependent T cell line, CTLL-2, indicating that the γ chain is functionally involved in transduction of cell growth signals mediated by IL-9. In contrast to the other cytokine receptors sharing the γ chain, no effect of TUGm2 on the affinity of IL-9 binding is seen.[84] TUGm3 precipitates IL-9-crosslinked γ chain, confirming the physical association among the γ chain, IL-9Rα and IL-9.

A novel cytokine, IL-15, has been initially detected in supernatants from a simian kidney epithelial cell line, CV-1/EBNA, as an IL-2-like T cell growth factor mediating proliferation of IL-2-dependent cells.[87] Unlike IL-2, IL-15 was then found to be produced by a wide variety of cells. Recombinant IL-15 induces proliferation of CTLL-2 cells, PHA-stimulated CD4$^+$ and CD8$^+$ T cells, and activates cytotoxic T cells and LAK cells, like IL-2. Interestingly, not only the γ chain but also IL-2Rβ is utilized as an essential component of the IL-15 receptor complex.[88] The IL-15 receptor requires an α chain for formation of the fully functional receptor, which is different than but structurally resembles the IL-2Rα.[89]

9.7 SIGNALS FROM THE IL-2 RECEPTOR

After the extracellular interaction between IL-2 and the receptors, besides internalization, the functional receptors mediate intracellular growth signals. Tyrosine phosphorylation occurs immediately after IL-2 binding, similar to the growth signals transduced by the growth factor receptors for PDGF, EGF and insulin, which carry intrinsic tyrosine kinase domains in the receptor molecules. IL-2-induces tyrosine phosphorylation of the γ chain as well as IL-2Rβ in vivo and in vitro.[90,91] No intrinsic tyrosine kinase domain is, however, found in the cytokine receptor molecules. Thus one may assume that a tyrosine kinase is associated directly or indirectly with the IL-2 receptor complex. Exploration of such a kinase demonstrated activation of $p56^{lck}$, a non-receptor type src-family tyrosine kinase, by IL-2.[92] In addition, $p56^{lck}$ is shown to be included in the immunoprecipitates with IL-2Rβ. Another non-receptor type tyrosine kinase, $p53/p56^{lyn}$, is also reported to be included in the immunoprecipitates with the β chain.[93] Nevertheless, it becomes obvious that non-receptor type tyrosine kinase(s) other than $p56^{lck}$ also participate in IL-2-induced signals, because cells which do not express $p56^{lck}$ show the similar tyrosine phosphorylation in response to IL-2.[91]

Currently, novel tyrosine kinases belonging to the family of Janus kinases (JAK) have been implicated in signal transduction initiated by several cytokines.[1,94] JAK family tyrosine kinases are shown to be functionally associated with the LIF receptor β chain, gp130 of the IL-6 receptor and receptors for Epo, IL-3, and granulocyte-colony stimulation factor (G-CSF).[95-98] In this context, we and others demonstrated that IL-2 induces activation of JAK1 and JAK3 but not JAK2 and Tyk2.[99-102] Immunoprecipitation experiments with anti-IL-2Rβ and anti-γ chain monoclonal antibodies demonstrate that JAK1 directly associates with IL-2Rβ, and JAK3 with the γ chain (Fig. 9.5). These association are independent of IL-2 binding to the IL-2 receptor. Interestingly, a mouse fibroblastoid 3T3 subline expressing the exogenous IL-2Rα, IL-2Rβ and γ chains respond to IL-2 for DNA synthesis only when transfected along with the JAK3 gene.[104] The same cell line without JAK3 has been shown to exhibit IL-2 responsiveness in terms of tyrosine phosphorylation and induction of proto-oncogenes, but not cell growth. Therefore, it is very likely that JAK3 activation is one of the essential events required for the IL-2-induced cell growth. Very recently, two autosomal recessive SCID patients with phenotypes of a lack of T cells and the presence of B cells, similar to XSCID, have been shown to carry mutated JAK3 genes.[105] These results clearly illustrate that γ chain-transduced signals essential for T cell development passes through JAK3.

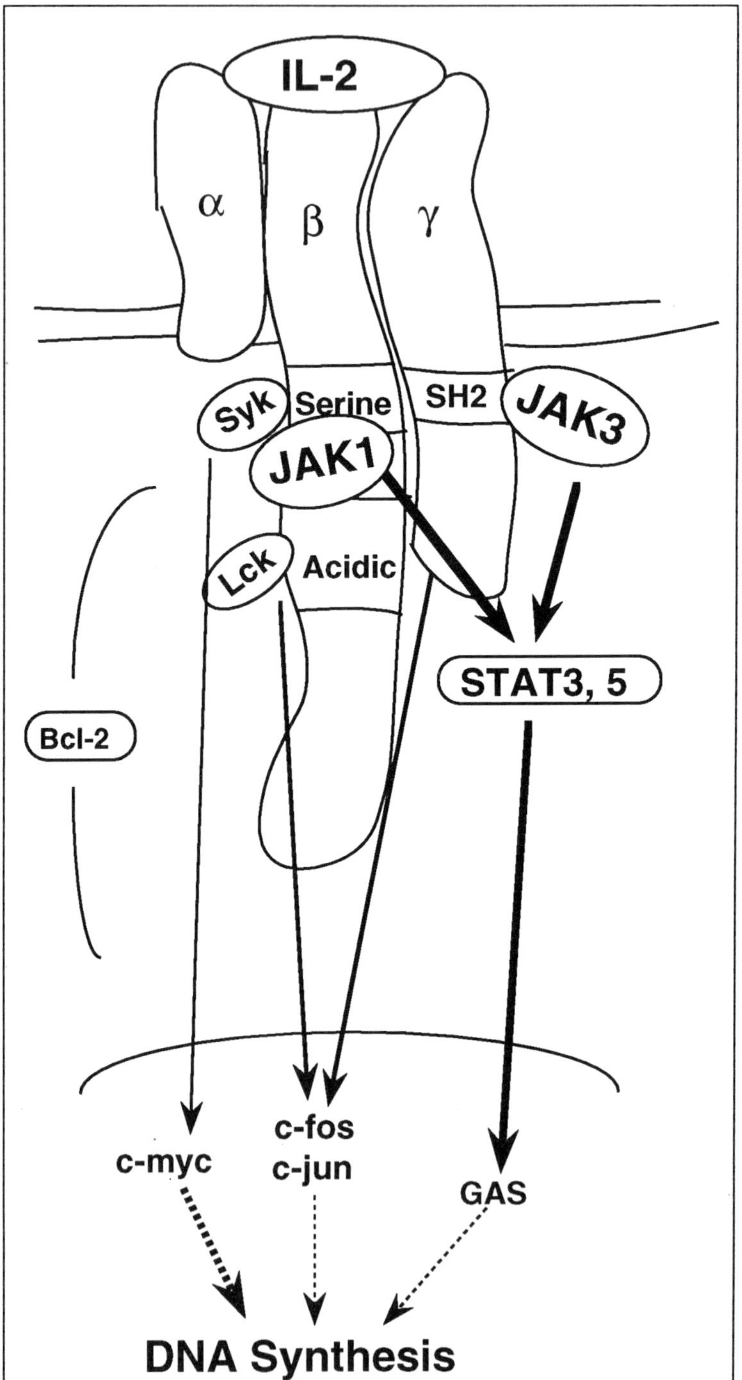

Fig. 9.5. Signal transduction pathway from the IL-2 receptor complex.

Recent studies have documented that the family of signal transducer and activator of transcription proteins (STAT) participate in the signal transducing pathways from cytokine receptors,[94] and are thought to play an important role downstream of the JAK family tyrosine kinases.[106] Seven STAT molecules have been identified so far. STAT3 and STAT5 were found to be involved in IL-2-mediated signal transduction.[103,107,108] IL-2 binding to the receptor induces activation of JAK1 and JAK3 associated with IL-2Rβ and γ chain, respectively. The activated JAK molecules phosphorylate IL-2Rβ and γ chain at tyrosine residues which may serve as binding sites for the STAT molecules. STAT3 and STAT5 recruited with the IL-2 receptor subunits are then tyrosine-phosphorylated presumably by the JAK tyrosine kinases, and again leave the receptor molecules to form homodimer or heterodimer through interaction between the SH2 domains and phosphorylated tyrosines of the STAT molecules. The STAT dimers move to the nucleus where they bind directly γ-activation sequences (GAS) in the regulatory regions of target genes (Fig. 9.5). This scheme is generalized for other cytokine-induced intracellular signals. The SH2 domains of STAT molecules selectively bind respective target sequences including phosphotyrosines of receptor subunits, possibly determining the specific correlation between STAT molecules and cytokines. Dissociation of activation of STAT and cell proliferation has been reported in the case of STAT5,[109] though further studies are required to support the finding. Constitutive activation of JAK3 has been seen in T cells transformed with HTLV-I.[110]

Reconstitution studies with mutants of IL-2Rβ and γ chain demonstrate that a γ chain mutant retaining the SH2-like region intact is able to induce c-myc gene induction in response to IL-2 but not c-fos and c-jun genes, and other mutants lacking a part or whole of the SH2-like region are no longer able to transduce any signal except for IL-2 internalization, indicating that the SH2-like region plays an important role in the IL-2-mediated intracellular signal transduction.[111] On the other hand, a mutant deleted for the acidic region (A-mutant) of IL-2Rβ shows IL-2-induced cell growth, whereas the other IL-2Rβ mutant lacking the serine-rich region (S-mutant) does not grow in response to IL-2. Only c-myc gene induction is detected in the A-mutant transfectants, while the S-mutant shows no significant induction of the c-jun, c-fos and c-myc genes.[112-114] These results suggest that the serine-rich region of IL-2Rβ plays a crucial role in signal transduction for IL-2-mediated cell growth. The acidic region deletion mutant fails to induce expression of c-fos and c-jun genes but retains the ability to induce expression of c-myc gene and cell growth. IL-2-induced c-myc induction always associates with tyrosine phosphorylation and IL-2-induced cell proliferation, while induction of the c-jun and c-fos genes appears to be less associated with cell proliferation,[99,115] indicating two distinct signaling pathways from the IL-2 receptor complex.

The cytoplasmic domain of IL-2Rβ also contains the regions known as box1 and box2, which are originally defined in the cytoplasmic domain of gp130, and are conserved among several cytokine receptors, i.e. those for G-CSF, Epo, and the common β chain.[116] Both box1 and box2 of IL-2Rβ are located proximal to the transmembrane domain, and box2 is included in the serine-rich region of IL-2Rβ. The membrane proximal domain including the box1 and serine-rich regions has been shown to be involved in cell growth signaling. Furthermore, using transfectants with various cytoplasmic deletion mutants of IL-2Rβ and γ chain, their association sites with JAK1 and JAK3 were dissected; the serine-rich region of IL-2Rβ and the transmembrane proximal region containing the SH2-like region of the γ chain participates in association with JAK1 and JAK3, respectively.[51,100,104]

Apart from tyrosine phosphorylation, activation of Raf-1, a cytoplasmic serine/threonine kinase, presumably participates in the growth signal pathway from the IL-2 receptor.[117,118] The Raf-1 activity is suggested to be regulated by phosphorylation of tyrosine residues in the Raf-1 molecule. This tyrosine phosphorylation occurs in response to IL-2 and may be directed by a tyrosine kinase associated with the IL-2 receptor. Another example of activation of a cytoplasmic serine/threonine kinase by IL-2 is S6 kinase which is suggested to be downstream of Raf-1.[119,120] The activation of this enzyme may be involved in the IL-2-induced signal cascades, followed by increase in protein synthesis for cell proliferation. Other biochemical changes associated with IL-2 activation are hydrolysis of glycosylphosphatidylinositol into myristylated diacylglycerol and inositolphosphate glycan,[121,122] both of which may act as second messengers, and activation of $p21^{ras}$,[123] whose signal may be delivered to the nucleus through the cascade including c-Raf-1 and S6 kinase. Activation of protein kinase C (PKC) by IL-2 has been controversial.[124-126] A recent study reveals that IL-2 apparently activates membrane-associated inactive PKC without its translocation from cytosol to membrane, suggesting that IL-2-induced activation of PKC may be implicated in IL-2-dependent survival of T cells.[127]

Related to Ras activation, IL-2 induces tyrosine phosphorylation of Shc and anti-Shc-directed immunoprecipitates from IL-2-stimulated cells contain a tyrosine kinase activity.[128] Furthermore, it is demonstrated that Shc is associated with IL-2Rβ. Tyrosine-phosphorylated Shc interacts with the SH2 domain of Grb2, another adapter protein, and Grb2 further binds to the Ras GDT/GTP exchange factor, SOS.[129,130] Thus, the cascade interaction of IL-2Rβ, Shc, Grb2, SOS and Ras, together with activation of tyrosine kinases, may be included in IL-2-induced signaling. IL-2 also induces activation of $p44^{erk\text{-}1}$ and $p42^{erk\text{-}2}$ in Con A-activated T cells.[131]

In addition to the nuclear proto-oncogenes, c-fos, c-jun, c-myc and c-myb, genes for cyclins and cdc2 kinase family, which function as cell cycle regulators, are transcriptionally induced in response to

IL-2.[132] Cell cycle progression is exerted by activation of cyclin-dependent kinases (CDKs) that are negatively regulated by G1 cyclin-CDK inhibitors. In IL-2-induced G1 progression to S phase, IL-2 is found to inactivate the Cdk2 inhibitor, probably $p27^{Kipl}$, which is included in the G1 cyclin-Cdk2 complex.[133]

IL-2-dependent cells generally die, presumably due to apoptosis, upon IL-2 deprivation, which induces downregulation of bcl-2 to undetectable levels in IL-2-dependent CTLL-2 cells. The re-expression of bcl-2 is induced by addition of IL-2, and protects the cells from apoptosis.[134]. Additionally, deregulated expression of bcl-2 prolongs the survival of CTLL-2 cells in the absence of IL-2. The interplay between IL-2 and expression of bcl-2 has important implications in the maintenance and growth of IL-2-dependent cells.

9.8 XSCID MODEL IN MOUSE

The mouse γ chain gene is on the X chromosome.[27] Disruption of the γ chain gene by gene targeting would be expected to establish a mouse model for human XSCID. We and two other groups have produced γ chain mutant mice.[135-137] Our mutant lacks the cytoplasmic domain, which mimics one of the mutations of human XSCID. This mutant is expressed on the cell surface, and would be expected to be functional in IL-2 binding but not to be functional in signal transduction. Other groups generated γ chain mutants which are completely null for its expression of the cell surface. All three mutant mice show substantially the same phenotypes of little or no NK cells and profound reduction of T and B cells. These phenotypes are close to those of human XSCID and quite similar to those of IL-7 dysfunction mice. They are, however, distinct from those of human XSCID in that production of T cells is not so profoundly inhibited but B cells are decreased in the γ chain mutant mice. The differences presumably suggest that the mechanisms by which mouse and human develop lymphocytes are different between two species. We observe with the mutant mice an increase in monocytes in the periphery and the atrophied thymus with thymocytes one-fiftieth less than the control (Fig. 9.6). In addition, our mutant mice show lymphoadenopathy of celiac lymph nodes and an increase in $CD34^+$ $c\text{-}kit^+$ $Sca1^+$ hematopoietic stem cell population. The lymphoadenopathy may reflect a predisposition to development of lymphomas often seen in patients with the primary and secondary immunodeficiencies. The γ chain mutant mice are deficient for humoral, T cell-mediated and NK cell-mediated immunities and are different from the SCID mouse strains previously found in terms of NK cell activity. They could presumably provide a SCID mouse model which would be used in experiments requiring more severely impaired immune function. The mutant mice would also be useful for development of gene therapy for human XSCID.

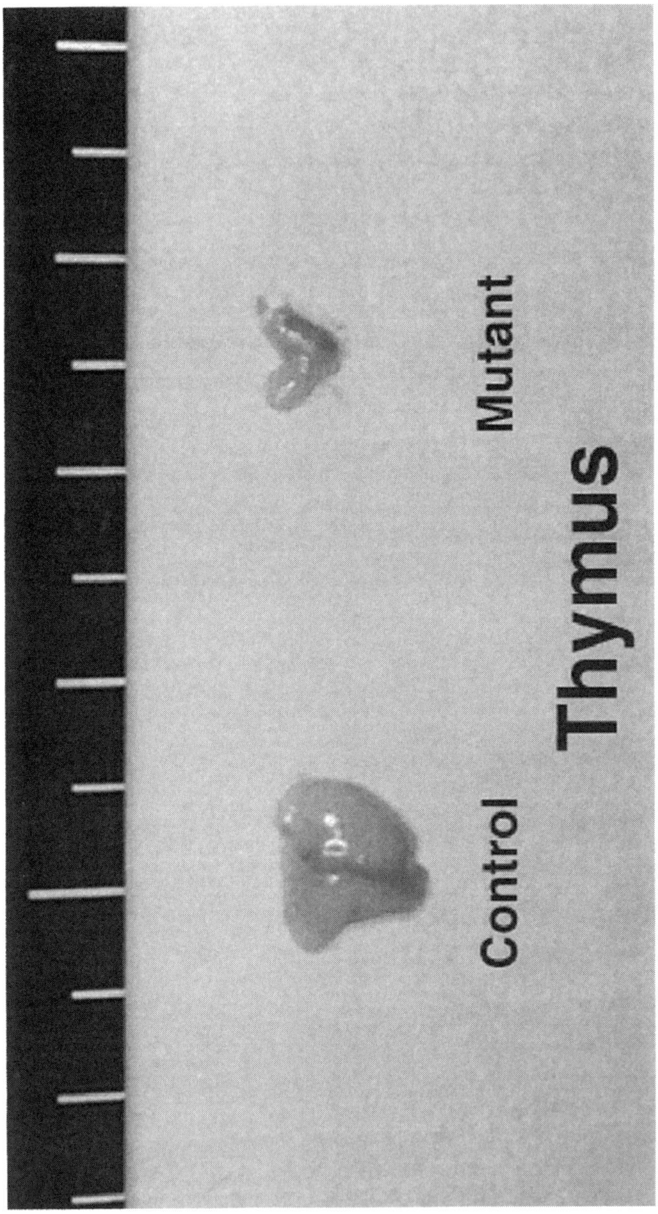

Fig. 9.6. Atrophied thymus in γ chain mutant mice. (Courtesy of Drs. K. Takahashi and M. Ikebe of Kumamoto University)

9.9 CONCLUDING REMARKS

Molecular cloning of the γ chain made it possible to gain great insight into its function. We would like to emphasize two issues. First, the γ chain is used as an essential subunit for the signal transduction in receptors for multiple cytokines, IL-2, IL-4, IL-7, IL-9 and IL-15, thus it is called the common γ chain. This may be the molecular basis of the pleiotropy of cytokines. Consequently, four members of the cytokine receptor superfamily have been found to be shared in multiple cytokine receptors so far. The second point is that dysfunctional mutations of the γ chain cause XSCID. Disruption of the γ chain gene in mice generates similar phenotypes to human XSCID, confirming that the γ chain plays a critical roles in lymphocyte development. This phenomenon suggests that the γ chain is one of the key molecules which determine cytoplasmic signals. Further studies will elucidate signaling pathways both common and specific to cytokines. We are now in a position to address the complexity of cytokine/cytokine receptor networks responsible for differentiation and proliferation of hematopoietic lineage cells. The detailed understanding of both receptor structure, ligand-receptor interaction and signal transduction events provides numerous points for therapeutic intervention. Based on this progress, it would be possible to develop new reagents which inhibit activation of immune responses by modification of the ligand receptor interaction and intracellular signalling, presumably resulting in alleviation of inflammation and regression of lymphomas and leukemias. Moreover, similar reagents would be useful in transplantation through subtle regulation of immunosuppression in patients. In addition, gene therapy for XSCID is now a credible option.

9.10 Acknowledgments

This work was supported in part by the grants from the Ministry of Education, Science, Sports and Culture of Japan, by the grant from Special Coordination Funds of the Science and Technology Agency of Japan, and by the Ciba-Geigy Foundation (Japan) for the Promotion of Science.

References

1. Taniguchi T. Cytokine signaling through nonreceptor protein tyrosine kinases. Science 1995; 268:251-5.
2. Bazan JF. A novel family of growth factor receptors: A common binding domain in the growth hormone, prolactin, erythropoietin and IL-6 receptors, and the p75 IL-2 receptor β-chain. Biochem Biophys Res Commun 1989; 164:788-95.
3. Cosman D, Lyman SD, Idzerda RL, Beckmann MP, Park LS, Goodwin RG, March CJ. A new cytokine receptor superfamily. TIBS 1990; 15:265-70.
4. Morgan DA, Ruscetti FW, Gallo R. Selective *in vitro* growth of T lymphocytes from normal human bone marrows. Science 1976; 193:1007-8.

5. Robb RJ, Greene WC, Rusk CM. Low and high affinity cellular receptors for interleukin 2. Implications for the level of Tac antigen. J Exp Med 1984; 160:1126-46.
6. Uchiyama T, Broder S, Waldmann TA. A monoclonal antibody (anti-Tac) reactive with activated and functionally mature human T cells. I. Production of anti-Tac monoclonal antibody and distribution of Tac(+) cells. J Immunol 1981; 126:1393-7.
7. Leonard WJ, Depper JM, Uchiyama T, Smith KA, Waldmann TA, Greene WC. A monoclonal antibody that appears to recognize the receptor for human T-cell growth factor; partial characterization of the receptor. Nature 1982; 300:267-9.
8. Bich-Thuy LT, Dukovich M, Peffer NJ, Fauci AS, Kehrl JH, Greene WC. Direct activation of human resting T cells by IL 2: The role of an IL 2 receptor distinct from the Tac protein. J Immunol 1987; 139:1550-6.
9. Tanaka T, Saiki O, Doi S, Negoro S, Kishimoto S. Interleukin 2 functions through novel interleukin 2 binding molecules in T cells. J Immunol 1988; 140:470-3.
10. Leonard WJ, Depper JM, Crabtree GR, Rudikoff S, Pumphrey J, Robb RJ, Krönke M, Svetlik PB, Peffer NJ, Waldmann TA, Green WC. Molecular cloning and expression of cDNAs for the human interleukin-2 receptor. Nature 1984; 311:626-31.
11. Nikaido T, Shimizu A, Ishida N, Sabe H, Teshigawara K, Maeda M, Uchiyama T, Yodoi J, Honjo T. Molecular cloning of cDNA encoding human interleukin-2 receptor. Nature 1984; 311:631-5.
12. Cosman D, Cerretti DP, Larsen A, Park L, March C, Dower S, Gillis S, Urdal D. Cloning, sequence and expression of human interleukin-2 receptor. Nature 1984; 312:768-71.
13. Tsudo M, Kozak RW, Goldman CK, Waldmann TA. Demonstration of a non-Tac peptide that binds interleukin 2: A potential participant in a multichain interleukin 2 receptor complex. Proc Natl Acad Sci USA 1986; 83:9694-8.
14. Sharon M, Klausner RD, Cullen BR, Chizzonite R, Leonard WJ. Novel interleukin-2 receptor subunit detected by cross-linking under high-affinity conditions. Science 1986; 234:859-63.
15. Teshigawara K, Wang H-M, Kato K, Smith KA. Interleukin 2 high-affinity receptor expression requires two distinct binding proteins. J Exp Med 1987; 165:223-38.
16. Dukovich M, Wano Y, Bich-Thuy LT, Katz P, Cullen BR, Kehrl JH, Greene WC. A second human interleukin-2 binding protein that may be a component of high-affinity interleukin-2 receptors. Nature 1987; 327:518-22.
17. Robb RJ, Rusk CM, Yodoi J, Greene WC. Interleukin 2 binding molecule distinct from the Tac protein: Analysis of its role in formation of high-affinity receptors. Proc Natl Acad Sci USA 1987; 84:2002-6.
18. Takeshita T, Goto Y, Tada K, Nagata K, Asao H, Sugamura K. Monoclonal antibody defining a molecule possibly identical to the p75 subunit of interleukin 2 receptor. J Exp Med 1989; 169:1323-32.

19. Tsudo M, Kitamura F, Miyasaka M. Characterization of the interleukin 2 receptor β chain using three distinct monoclonal antibodies. Proc Natl Acad Sci USA 1989; 86:1982-6.
20. Hatakeyama M, Tsudo M, Minamoto S, Kono T, Doi T, Miyata T, Miyasaka M, Taniguchi T. Interleukin-2 receptor β chain gene: Generation of three receptor forms by cloned human α and β chain cDNA's. Science 1989; 244:551-6.
21. Minamoto S, Mori H, Hatakeyama M, Kono T, Doi T, Ide T, Uede T, Taniguchi T. Characterization of the heterodimeric complex of human IL-2 receptor α·β chains reconstituted in a mouse fibroblast cell line, L929. J Immunol 1990; 145:2177-82.
22. Takeshita T, Ohtani K, Asao H, Kumaki S, Nakamura M, Sugamura K. An associated molecule, p64, with IL-2 receptor β chain. Its possible involvement in the formation of the functional intermediate-affinity IL-2 receptor complex. J Immunol 1992; 148:2154-8.
23. Takeshita T, Asao H, Ohtani K, Ishii N, Kumaki S, Tanaka N, Munakata H, Nakamura M, Sugamura K. Cloning of the γ chain of the human IL-2 receptor. Science 1992; 257:379-82.
24. Koch C A, Anderson D, Moran MF, Ellis C, Pawson T. SH2 and SH3 domains: elements that control interactions of cytoplasmic signaling proteins. Science 1991; 252:668-74.
25. Mayer BJ, Hanafusa H. Mutagenic analysis of the v-*crk* oncogene: Requirement for SH2 and SH3 domains and correlation between increased cellular phosphotyrosine and transformation. J Virol 1990; 64:3581-9.
26. Kumaki S, Kondo M, Takeshita T, Asao H, Nakamura M, Sugamura K. Cloning of the mouse interleukin 2 receptor γ chain: demonstration of functional differences between the mouse and human receptors. Biochem Biophys Res Commun 1993; 193:356-63.
27. Cao X, Kozak CA, Liu Y, Noguchi M, O'Conell E, Leonard WJ. Characterization of cDNAs encoding the murine interleukin 2 receptor (IL-2R) γ chain: chromosomal mapping and tissue specificity of IL-2R γ chain expression. Proc Natl Acad Sci USA 1993; 90:8464-8.
28. Matsuoka M, Takeshita T, Ishii N, Nakamura M, Ohkubo T, Sugamura K. Kinetic study of interleukin-2 binding on the reconstituted interleukin-2 receptor complexes including the human γ chain. Eur J Immunol 1993; 23:2472-6.
29. Shimizu A, Kondo S, Takeda S, Yodoi J, Ishida N, Sabe H, Osawa H, Diamantstein T, Honjo T. Nucleotide sequence of mouse IL-2 receptor cDNA and its comparison with the human IL-2 receptor sequence. Nucleic Acids Res 1985; 13:1505-16.
30. Miller J, Malek TR, Leonard WJ, Greene WC, Shevach EM, Germain RN. Nucleotide sequence and expression of a mouse interleukin 2 receptor cDNA. J Immunol 1985; 134:4212-7.
31. Kono T, Doi T, Yamada G, Hatakeyama M, Minamoto S, Tsudo M, Miyasaka M, Miyata T, Taniguchi T. Murine interleukin 2 receptor

β chain: dysregulated gene expression in lymphoma line EL-4 caused by a promoter insertion. Proc Natl Acad Sci USA 1990; 87:1806-10.
32. Nemoto T, Takeshita T, Ishii N, Kondo M, Higuchi M, Satomi S, Nakamura M, Mori S, Sugamura K. Difference in interleukin 2 (IL-2) receptor system between human and mouse: Essential requirement of the α chain for formation of the functional mouse IL-2 receptor. Eur J Immunol 1995; 25: 3001-5.
33. Siegel JP, Sharon M, Smith PL, Leonard WJ. The interleukin β chain (p70): role in mediating signals for LAK, NK, and proliferative activities. Science 1987; 238:75-8.
34. Phillips JH, Takeshita T, Sugamura K, Lanier LL. Activation of natural killer cells via the p75 interleukin 2 receptor. J Exp Med 1989; 170:291-6.
35. Yagita H, Nakata M, Azuma A, Nitta T, Takeshita T, Sugamura K, Okumura K. Activation of peripheral blood T cells via the p75 interleukin 2 receptor. J Exp Med 1989; 170:1445-50.
36. Ishii N, Takeshita T, Kimura Y, Tada K, Kondo M, Nakamura M, Sugamura K, Expression of the IL-2 receptor γ chain on various populations in human peripheral blood. Int Immunol 1994; 6:1273-7.
37. Ohbo K, Takasawa N, Ishii N, Tanaka N, Nakamura M, Sugamura K. Functional analysis of the human interleukin 2 receptor γ chain gene promoter. J Biol Chem 1995; 270:7479-86.
38. Kondo M, Ohashi Y, Tada K, Nakamura M, Sugamura K. Expression of the mouse interleukin-2 receptor γ chain in various cell populations of the thymus and spleen. Eur J Immunol 1994; 24:2026-30.
39. Noguchi M, Adelstein S, Cao X, Leonard WJ. Characterization of the human interleukin-2 receptor γ chain gene. J Biol Chem 1993; 268:13601-8.
40. Yoshida M, Seiki M. Recent advances in the molecular biology of HTLV-I: *trans*-activation of viral and cellular genes. Annu Rev Immunol 1987; 5:541-59.
41. Inoue J, Seiki M, Taniguchi T, Tsuru S, Yoshida M. Induction of interleukin 2 receptor gene expression by p40x encoded by human T-cell leukemia virus type I. EMBO J 1986; 5:2883-8.
42. Bosco MC, Espinoza-Delgado I, Schwabe M, Gusella GL, Longo DL, Sugamura K, Varesio L. Regulation by interleukin-2 (IL-2) and interferon γ of IL-2 receptor γ chain gene expression in human monocytes. Blood 1994; 83:2995-3002.
43. Noguchi M, Yi H, Rosenblatt HM, Filipovich AH, Adelstein S, Modi WS, McBride OW, Leonard WJ. Interleukin-2 receptor γ chain mutation results in X-linked severe combined immunodeficiency in humans. Cell 1993; 73:147-57.
44. Puck JM, Deschénes SM, Porter JC, Dutra AS, Brown CJ, Willard HF, Henthorn PS. The interleukin-2 receptor γ chain maps to Xq13.1 and is mutated in X-linked severe combined immunodeficiency, SCIDX1. Hum Mol Genet 1993; 2:1099-104.
45. Ishii N, Asao H, Kimura Y, Takeshita T, Nakamura M, Tsuchiya S, Konno T, Maeda M, Uchiyama T, Sugamura K. Impairment of ligand

binding and growth signaling of mutant IL-2 receptor γ-chains in patients with X-linked severe combined immunodeficiency. J Immunol 1994; 153:1310-7.
46. DiSanto JP, Dautry Varsat A, Certain S, Fischer A, de Saint Basile G. Interleukin-2 (IL-2) receptor γ chain mutations in X-linked severe combined immunodeficiency disease result in the loss of high-affinity IL-2 receptor binding. Eur J Immunol 1994; 24:475-9.
47. Markiewicz S, Subtil A, Dautry-Varsat A, Fischer A, de Saint Basile G. Detection of three nonsense mutations and one missense mutation in the interleukin-2 receptor γ chain gene in SCIDX1 that differently affect the mRNA processing. Genomics 1994; 21:291-3.
48. Conley ME. Molecular approaches to analysis of X-linked immunodeficiencies. Annu Rev Immunol 1992; 10:215-38.
49. O'Reilly RJ, Dupont B, Pahwa S, Grimes E, Smithwick EM, Pahwa R, Schwartz S, Hansen JA, Siegal FP, Sorell M, Svejgaard A, Jersild C, Thomsen M, Platz P, L'Esperance P, Good RA. Reconstitution in severe combined immunodeficiency by transplantation of marrow from an unrelated donor. N Engl J Med 1977; 297:1311-8.
50. Buckley RH, Schiff SE, Sampson HA, Schiff RI, Markert ML, Knutsen AP, Hershfield MS, Huang AT, Mickey GH, Ward FE. Development of immunity in human severe primary T cell deficiency following haploidentical bone marrow stem cell transplantation. J Immunol 1986; 136:2398-407.
51. Russell SM, Johnston JA, Noguchi M, Kawamura M, Bacon CM, Friendmann M, Berg M, McVicar DW, Witthuhn BA, Silvennoinen O, Goldman AS, Schmalstieg FC, Ihle JN, O'Shea JJ, Leonard WJ. Interaction of IL2Rβ and γ_c chains with Jak1 and Jak3: implications for XSCID and XCID. Science 1994; 266:1042-5.
52. Seeger RC, Robins RA, Stevens RH, Klein RB, Waldman DJ, Zeltzer PM, Kessler SW. Severe combined immunodeficiency with B lymphocytes: *in vitro* correction of defective immunoglobulin production by addition of normal T lymphocytes. Clin Exp Immunol 1976; 26:1-10.
53. Conley ME, Lavoie A, Briggs C, Brown P, Guerra C, Puck JM. Nonrandom X chromosome inactivation in B cells from carriers of X chromosome-linked severe combined immunodeficiency. Proc Natl Acad Sci USA 1988; 85:3090-4.
54. Puck JM, Nussbaum RL, Conley ME. Carrier detection in X-linked severe combined immunodeficiency based on patterns of X chromosome inactivation. J Clin Invest 1987; 79:1395-400.
55. Goodship J, Malcolm S, Lau YL, Pembrey ME, Levinsky RJ. Use of X chromosome inactivation analysis to establish carrier status for X-linked severe combined immunodeficiency. Lancet 1988; 1:729-32.
56. DiSanto JP, Rieux-Laucat F, Dautry-Varsat A, Fischer A, de Saint Basile G. Defective human interleukin 2 receptor γ chain in an atypical X chromosome-linked severe combined immunodeficiency with peripheral T cells. Proc Natl Acad Sci USA 1994; 91:9466-70.
57. Brooks EG, Schmalstieg FC, Wirt DP, Rosenblatt HM, Adkins LT,

Lookingbill D P, Rudloff HE, Rakusan TA, Goldman AS. A novel X-linked combined immunodeficiency disease. J Clin Invest 1990; 86:1623-31.
58. Henthorn PS, Somberg RL, Fimiani VM, Puck JM, Patterson DF, Felsburg PJ. IL-2R γ gene microdeletion demonstrates that canine X-linked severe combined immunodeficiency is a homologue of the human disease. Genomics 1994; 23:69-74.
59. Somberg RL, Robinson JP, Felsburg PJ. T lymphocyte development and function in dogs with X-linked severe combined immunodeficiency. J Immunol 1994; 153:4006-15.
60. Pahwa R, Chatila T, Pahwa S, Paradise C, Day NK, Geha R, Schwartz SA, Slade H, Oyaizu N, Good RA. Recombinant interleukin 2 therapy in severe combined immunodeficiency disease. Proc Natl Acad Sci USA 1989; 86:5069-73.
61. Weinberg K, Parkman R. Severe combined immunodeficiency due to a specific defect in the production of interleukin-2. N Engl J Med 1990; 322:1718-23.
62. DiSanto JP, Keever CA, Small TN, Nicols GL, O'Reilly RJ, Flomenberg N. Absence of interleukin 2 production in a severe combined immunodeficiency disease syndrome with T cells. J Exp Med 1990; 171:1697-704.
63. Schorle H, Holtschke T, Hünig T, Schimpl A, Horak I. Development and function of T cells in mice rendered interleukin-2 deficient by gene targeting. Nature 1991; 352:621-21.
64. Sadlack B, Merz H, Schorle H, Schimpl A, Feller AC, Horak I. Ulcerative colitis-like disease in mice with a disrupted interleukin-2 gene. Cell 1993; 75:253-61.
65. Kündig T M, Schorle H, Bachmann M F, Hengartner H, Zinkernagel R M, Horak I. Immune responses in interleukin-2-deficient mice. Science 1993; 262:1059-61.
66. Kondo M, Takeshita T, Ishii N, Nakamura M, Watanabe S, Arai K, Sugamura K. Sharing of the interleukin-2 (IL-2) receptor γ chain between receptors for IL-2 and IL-4. Science 1993; 262:1874-7.
67. Mosley B, Beckmann MP, March C J, Idzerda RL, Gimpel SD, VandenBos T, Friend D, Alpert A, Anderson D, Jackson J, Wignall JM, Smith C, Gallis B, Sims JE, Urdal D, Widmer MB, Cosman D,. Park LS. The murine interleukin-4 receptor: Molecular cloning and characterization of secreted and membrane bound forms. Cell 1989; 59:335-48.
68. Harada N, Castle BE, Gornan DM, Itoh N, Schreurs J, Barrett RL, Howard M, Miyajima A. Expression cloning a cDNA encoding the murine interleukin 4 receptor based on ligand binding. Proc Natl Acad USA 1990; 87:857-61.
69. Russell SM, AD, Harada N, Nakamura Y, Noguchi M, Leland P, Friedmann MC, Miyajima A, Puri RK, Paul WE, Leonard WJ. Interleukin-2 receptor γ chain: a functional component of the interleukin-4 receptor. Science 1993; 262:1880-3.
70. Mosmann TR, Cherwinski H, Bond MW, Giedlin MA, Coffman RL.

Two types of murine helper T cell clone. I. Definition according to profiles of lymphokine activities and secreted proteins. J Immunol 1986; 136:2348-57.

71. Hu-Li J, Shevach EM, Mizoguchi J, Ohara J, Mosmann T, Paul WE. B cell stimulatory factor 1 (interleukin 4) is a potent costimulant for normal resting T lymphocytes. J Exp Med 1987; 165:157-72.
72. Kühn R, Rajewsky K, Müller W. Generation and analysis of interleukin-4 deficient mice. Science 1991; 254:707-10.
73. Aversa G, Punnonen J, Cocks BG, Malefyt RW, Vega F, Zurawski SM, Zurawski G, Vries JE. An interleukin 4 (IL-4) mutant protein inhibits both IL-4 or IL-13-induced human immunoglobulin G4 (IgG4) and IgE synthesis and B cell proliferation: support for a common component shared by IL-4 and IL-13 receptors. J Exp Med 1993; 178:2213-8.
74. Zurawski S M, Vega F, Huyghe B, Zurawski G. Receptors for interleukin-13 and interleukin-4 are complex and share a novel component that functions in signal transduction. EMBO J 1993; 12:2663-70.
75. Kondo M, Takeshita T, Higuchi M, Nakamura M, Sudo T, Nishikawa S, Sugamura K. Functional participation of the IL-2 receptor γ chain in IL-7 receptor complexes. Science 1994; 263:1453-4.
76. Noguchi M, Nakamura Y, Russel SM, Ziegler SF, Tsang M, Cao X, Leonard WJ. Interleukin-2 receptor γ chain: a functional component of the interleukin-7 receptor. Science 1993; 262:1877-80.
77. Grabstein KH, Waldschmidt TJ, Finkelman FD, Hess BW, Alpert AR, Boiani NE, Namen AE, Morrissey PJ. Inhibition of murine B and T lymphopoiesis *in vivo* by an anti-interleukin 7 monoclonal antibody. J Exp Med 1993; 178:257-64.
78. Sudo T, Nishikawa S, Ohno N, Akiyama N, Tamakoshi M, Yoshida H, Nishikawa S. Expression and function of the interleukin 7 receptor in murine lymphocytes. Proc Natl Acad Sci USA 1993; 90:9125-9.
79. Peschon JJ, Morrissey PJ, Grabstein KH, Ramsdell FJ, Maraskovsky E, Gliniak BC, Park LS, Ziegler SF, Williams DE, Ware CW, Meyer JD, Davison BL. Early lymphocyte expansion is severely impaired in interleukin 7 receptor-deficient mice. J Exp Med 1994; 180:1955-60.
80. Hozumi K, Kondo M, Nozaki H, Kobori A, Nishimura T, Nishikawa S, Sugamura K, Habu S. Implication of the common γ chain of the IL-7 receptor in intrathymic development of pro-T cells. Int Immunol 1994; 6:1451-4.
81. Namen AE, Schmierer AE, March CJ, Overell RW, Park LS, Urdal DL, Mochizuki DY. B cell precursor growth-promoting activity purification and characterization of a growth factor active on lymphocyte precursors. J Exp Med 1988; 167:988-1002.
82. Namen AE, Lupton S, Hjerrild K, Wignall J, Mochizuki DY, Schmierer A, Mosley B, March CJ, Urdal D, Gillis S, Cosman D, Goodwin RG. Stimulation of B-cell progenitors by cloned murine interleukin-7. Nature 1988; 333:571-3.
83. von Freeden-Jeffry U, Vieira P, Lucian AL, McNeil T, Burdach GES,

Murray R. Lymphopenia in interleukin (IL)-7 gene-deleted mice identifies IL-7 as a nonredundant cytokine. 1995; J Exp Med 181:1519-26.
84. Kimura Y, Takeshita T, Kondo M, Ishii N, Nakamura M, Van Snick J, Sugamura K. Sharing of the IL-2 receptor γ chain with the functional IL-9 receptor complex. Int Immunol 1995; 7:115-20.
85. Suda T, Zlotnik A. IL-7 maintains the cell precursor potential of $CD3^-CD4^-CD8^-$ thymocytes J Immunol 1990; 146:3068-73.
86. Houssiau FA, Renauld JC, Stevens M, Lehmann F, Lethe B, Coulie PG, Van SJ. Human T cell lines and clones respond to IL-9. J Immunol 1993; 150:2634-40.
87. Grabstein KH, Eisenman J, Shanebeck K, Rauch C, Srinivasan S, Fung V, Beers C, Richardson J, Schoenborn MA, Ahdieh M. Cloning of a T cell growth factor that interacts with the β chain of the interleukin-2 receptor. Science 1994; 264:965-8.
88. Giri JG, Ahdieh M, Eisenman J, Shanebeck K, Grabstein K, Kumaki S, Namen A, Park LS, Cosman D, Anderson D. Utilization of the β and γ chains of the IL-2 receptor by the novel cytokine IL-15. EMBO J 1994; 13:2822-30.
89. Giri JG, Kumaki S., Ahdieh M, F. D. J, Loomis A, Shanebeck K, DuBose R, Cosman D, Park LS, Anderson DM. Identification and cloning of a novel IL-15 binding protein that is structurally related to the α-chain of the IL-2 receptor. EMBO J 1995; 14:3654-63.
90. Asao H, Takeshita T, Nakamura M, Nagata K, Sugamura K. Interleukin 2 (IL-2)-induced tyrosine phosphorylation of IL-2 receptor p75. J Exp Med 1990; 171:637-44
91. Asao H, Kumaki S, Takeshita T, Nakamura M, Sugamura K. IL-2-dependent *in vivo* and *in vitro* tyrosine phosphorylation of IL-2 receptor γ chain. FEBS Letters 1992; 304:141-5
92. Horak ID, Gress RE, Lucas PJ, Horak EM, Waldmann TA, Bolen JB. T-lymphocyte interleukin 2-dependent tyrosine protein kinase signal transduction involves the activation of $p56^{lck}$. Proc Natl Acad Sci USA 1991; 88:1996-2000.
93. Torigoe T, Saragovi HU, Reed JC. Interleukin 2 regulates the activity of the lyn protein-tyrosine kinase in a B-cell line. Proc Natl Acad Sci USA 1992; 89:2674-8.
94. Ihle JN. Cytokine receptor signaling. Nature 1995; 377:591-4.
95. Witthuhn BA, Quelle FW, Silvennoinen O, Yi T, Tang B, Miura O, Ihle JN. JAK2 associates with the erythropoietin receptor and is tyrosine phosphorylated and activated following stimulation with erythropoietin. Cell 1993; 74:227-36.
96. Silvennoinen O, Witthuhn B A, Quelle FW, Cleveland JL, Yi T, Ihle JN. Structure of murine Jak2 protein-tyrosine kinase and its role in interleukin 3 signal transduction. Proc Natl Acad Sci USA 1993; 90:8429-33.
97. Lütticken C, Wegenka UM, Yuan J, Buschmann J, Schindler C, Ziemiecki A, Harpur AG, Wilks AF, Yasukawa K, Taga T, Kishimoto T, Barbieri

G, Pellegrini S, Sendtner M, Heinrich PC, Horn F. Association of transcription factor APRF and protein kinase Jak1 with the interleukin-6 signal transducer gp130. Science 1994; 263:89-92.

98. Stahl N, Boulton TG, Farruggella T, Ip NY, Davis S, Witthuhn BA, Quelle FW, Silvennoinen O, Barbieri G, Pellegrini S, Ihle JN, Yancopoulos GD. Association and activation of Jak-Tyk kinases by CNTF-LIF-OSM-IL-6 β receptor components. Science 1994; 263:92-5.
99. Tanaka N, Asao H, Ohbo K, Ishii N, Takeshita T, Nakamura M, Sasaki H, Sugamura K. Physical association of JAK1 and JAK2 tyrosine kinases with the interleukin 2 receptor β and γ chains. Proc Natl Acad Sci USA 1994; 91:7271-5.
100. Asao H, Tanaka N, Ishii N, Higuchi M, Takeshita T, Nakamura M, Shirasawa T, Sugamura K. Interleukin 2-induced activation of Jak3: possible involvement in signal transduction for c-myc induction and cell proliferation. FEBS Letters 1994; 351:201-6.
101. Johnston JA, Kawamura M, Kirken RA, Chen YQ, Blake TB, Shibuya K, Ortaldo JR, McVicar DW, O'Shea JJ. Phosphorylation and activation of the Jak-3 Janus kinase in response to interleukin-2. Nature 1994; 370:151-3.
102. Kirken RA, Rui H, Malabarba MG, Farrar WL. Identification of the interleukin-2 receptor-associated tyrosine kinase p116 as novel leukocyte-specific Janus kinase. J Biol Chem 1994; 269:19136-41.
103. Beadling C, Guschin D, Witthuhn BA, Ziemiecki A, Ihle JN, Kerr IM, Cantrell DA. Activation of JAK kinases and STAT proteins by interleukin-2 and interferon α, but not the T cell antigen receptor, in human T lymphocytes. EMBO J 1994; 13:5605-15.
104. Miyazaki T, Kawahara A, Fujii H, Nakagawa Y, Minami Y, Liu ZJ, Oishi I, Silvennoinen O, Witthuhn BA, Ihle JN, Taniguchi T. Functional activation of Jak1 and Jak3 by selective association with IL-2 receptor subunits. Science 1994; 266:1045-7.
105. Macchi P, Villa A, Gilliani S, Sacco MG, Frattini A, Porta F, Ugazio AG, Johnston JA, Candotti F, O'Shea JJ, Vezzoni P, Notarangelo LD. Mutations of Jak-3 gene in patients with autosomal severe combined immune deficiency (SCID). Nature 1995; 377:65-8.
106. Darnell JJ, Kerr IM, Stark GR. Jak-STAT pathways and transcriptional activation in response to IFNs and other extracellular signaling proteins. Science 1994; 264:1415-21.
107. Lamb P, Kessler LV, Suto C, Levy DE, Seidel HM, Stein RB, Rosen J. Rapid activation of proteins that interact with the interferon γ activation site in response to multiple cytokines. Blood 1994; 83:2063-71.
108. Gilmour KC, Reich NC. Receptor to nucleus signaling by prolactin and interleukin 2 via activation of latent DNA-binding factors. Proc Natl Acad Sci USA 1994; 91:6850-4.
109. Fujii H, Nakagawa Y, Schindler U, Kawahara A, Mori H, Gouilleux F, Groner B, Ihle JN, Minami Y, Miyazaki T, Taniguchi T. Activation of Stat5 by interleukin 2 requires a carboxy-terminal region of the interleukin

2 receptor β chain but is not essential for the proliferative signal transmission. Proc Natl Acad Sci USA 1995; 92:5482-6.
110. Migone TS, Lin JX, Cereseto A, Mulloy JC, O'Shear JJ, Franchini G, Leonard WJ. Constitutively activated Jak-STAT pathway in T cells transformed with HTLV-I. Science 1995; 269:79-81.
111. Asao H, Takeshita T, Ishii N, Kumaki S, Nakamura M, Sugamura K. Reconstitution of the functional IL-2 receptor complexes on fibroblastoid cells: Involvement of the cytoplasmic domain of the γ chain in two distinct signaling pathways. Proc Natl Acad Sci USA 1993; 90:4127-31.
112. Hatakeyama M, Mori H, Doi T, Taniguchi T. A restricted cytoplasmic region of IL-2 receptor β chain is essential for growth signal transduction but not for ligand binding and internalization. Cell 1989; 59:837-45.
113. Hatakeyama M, Kawahara A, Mori H, Shibuya H, Taniguchi T. c-*fos* gene induction by interleukin 2: identification of the critical cytoplasmic regions within the interleukin 2 receptor β chain. Proc Natl Acad Sci USA 1992; 89:2022-6.
114. Satoh T, Minami Y, Kono T, Yamada K, Kawahara A, Taniguchi T, Kaziro Y. Interleukin 2-induced activation of Ras requires two domains of interleukin 2 receptor β subunit, the essential region for growth stimulation and Lck-binding domain. J Biol Chem 1992; 267:25423-7.
115. Asao, H., Kumaki, S., Takeshita, T., Nakamura, M. and Sugamura, K. IL-2-dependent *in vivo* and *in vitro* tyrosine phosphorylation of IL-2 receptor γ chain. FEBS Letters 1992; 304:141-5.
116. Murakami M, Narazaki M, Hibi M, Yawata H, Yasukawa K, Hamaguchi M, Taga T, Kishimoto T. Critical cytoplasmic region of the interleukin 6 signal transducer gp130 is conserved in the cytokine receptor family. Proc. Natl. Acad. Sci. USA 1991; 88:11349-53.
117. Turner B, Rapp U, App H, Greene M, Dobashi K, Reed J. Interleukin 2 induces tyrosine phosphorylation and activation of p72-74 Raf-1 kinase in a T-cell line. Proc Natl Acad Sci USA 1991; 88:1227-31.
118. Zmuidzinas A, Mamon HJ, Roberts TM, Smith KA. Interleukin-2-triggered Raf-1 expression, phosphorylation, and associated kinase activity increase through G_1 and S in CD3-stimulated primary human T cells. Mol Cell Biol 1991; 11:2794-803.
119. Evans SW, Farrar WL. Interleukin 2 and diacylglycerol stimulate phosphorylation of 40 S ribosomal S6 protein. J Biol Chem 1987; 262:4624-30.
120. Kuo CJ, Chung J, Fiorentino DF, Flanagan WM, Blenis J, Crabtree GR. Rapamycin selectively inhibits interleukin-2 activation of p70 S6 kinase. Nature 1992; 358:70-3.
121. Merida I, Pratt JC, Gaulton GN. Regulation of interleukin 2-dependent growth responses by glycosylphosphatidylinositol molecules. Proc Natl Acad Sci USA 1990; 87:9421-5.
122. Eardley DD, Koshland ME. Glycosylphosphatidylinositol: A candidate system for interleukin-2 signal transduction. Science 1991; 251:78-81.
123. Satoh T, Nakafuku M, Miyajima A, Kaziro Y. Involvement of ras p21 protein in signal-transduction pathways from interleukin 2, interleukin 3,

and granulocyte/macrophage colony-stimulating factor, but not from interleukin 4. Proc Natl Acad Sci USA 1991; 88:3314-8.

124. Farrar WL, Anderson WB. Interleukin-2 stimulates association of protein kinase C with plasma membrane. Nature 1985; 315:233-5.
125. Mills GB, Girard P, Grinstein S, Gelfand EW. Interleukin-2 induces proliferation of T lymphocyte mutants lacking protein kinase C. Cell 1988; 55:91-100.
126. Valge VE, Wong JGP, Datlof BM, Sinskey AJ, Rao A. Protein kinase C is required for responses to T cell receptor ligands but not to interleukin-2 in T cells. Cell 1988; 55:101-12.
127. Lu Y, Tremblay R, Jouishomme H, Chakravarthy B, Durkin JP. Evidence that the activation of an inactive pool of membrane-associated protein kinase C is linked to the IL-2-dependent survival of T lymphocytes. J Immunol 1994; 153:1495-504.
128. Burns LA, Karnitz LM, Sutor SL, Abraham RT. Interleukin-2-induced tyrosine phosphorylation of $p52^{shc}$ in T lymphocytes. J Biol Chem 1993; 268:17659-61.
129. Ravichandran KS, Burakoff SJ. The adapter protein shc interacts with the interleukin-2 (IL-2) receptor upon IL-2 stimulation. J Biol Chem 1994; 269:1599-602.
130. Zhu X, Suen KL, Barbacid M, Bolen JB, Fargnoli J. Interleukin-2-induced tyrosine phosphorylation of Shc proteins correlates with factor-dependent T cell proliferation. J Biol Chem 1994; 269:5518-22.
131. Watts JD, Welham MJ, Kalt L, Schrader JW, Aebersold R. IL-2 stimulation of T lymphocytes induces sequential activation of mitogen-activated protein kinases and phosphorylation of $p56^{lck}$ at serine-59. J Immunol 1993; 151:6862-71.
132. Shibuya H, Yoneyama M, Ninomiya-Tsuji J, Matsumoto K, Taniguchi T. IL-2 and EGF receptors stimulate the hematopoietic cell cycle via different signaling pathways: demonstration of a novel role for c-myc. Cell 1992; 70:57-67.
133. Firpo EJ, Koff A, Solomon MJ, Roberts JM. Inactivation of a Cdk2 inhibitor during interleukin 2-induced proliferation of human T lymphocytes. Mol Cell Biol 1994; 14:4889-901.
134. Deng G, Podack ER. Suppression of apoptosis in a cytotoxic T-cell line by interleukin 2-mediated gene transcription and deregulated expression of the protooncogene *bcl-2*. Proc Natl Acad Sci USA 1993; 90:2189-93.
135. DiSanto JP, Müller W, Guy-Grand D, Fischer A, Rajewsky K. Lymphoid development in mice with a targeted deletion of the interleukin 2 receptor γ chain. Proc Natl Acad Sci USA 1995; 92: 377-81.
136. Cao X, Shores EW, Hu-Li J, Anver MR, Kelsall BL, Russell SM, Drago J, Noguchi M, Grinberg A, Bloom ET, Paul WE, Katz SI, Love PE, Leonard WJ. Defective lymphoid development in mice lacking expression of the common cytokine receptor γ chain. Immunity 1995; 2:223-38.

137. Ohbo K, Suda T, Hashiyama M, Mantani A, Ikebe M, Miyakawa K, Moriyama M, Nakamura M, Katsuki M, Takahashi K, Yamamura K, Sugamura K. Modulation of hematopoiesis in mouse with a truncated mutant of the interleukin 2 receptor γ chain. Blood 1996; 87: 956-67.

CHAPTER 10

Complement Receptors and Their Therapeutic Applications

Danuta E. Mossakowska and Richard A.G. Smith

10.1 INTRODUCTION

The complement system is a collection of some 25 plasma and membrane-bound proteins which act to 'complement' immunoglobulins in the function of humoral immunity. In this role, complement mediates four main types of biological activity essential to the detection and neutralization of foreign organisms, tissues and proteins. These are binding and transport, labeling and opsonization, focused inflammation and cytolysis. Despite the critical importance of these processes for the survival of higher organisms, the complement system has been (in B.P. Morgan's words) "through no fault of its own, unloved."[1] Such unpopularity–the product of apparent complexity, methodological difficulty and terminological confusion–is no longer justified. The application of molecular biology to an impressive body of biochemical knowledge about complement has clarified and rationalized the system to the point where molecular design of therapeutic agents has become possible. The purpose of this chapter is to survey our current knowledge of the complement receptors and to outline their impact on drug design.

Complement activation is conventionally divided into two arms, termed the classical and alternative pathways, which are depicted in a simplified form in Figure 10.1. For more details of the activation steps and the properties of the pathway components, the reader is referred to several excellent recent books and reviews.[1-4] The pathways of complement activation rely on the amplification effect of enzyme cascades, particularly the C3 convertases (which cleave plasma C3 to C3b) and

Recombinant Cell Surface Receptors: Focal Point for Therapeutic Intervention, edited by Michael J. Browne.

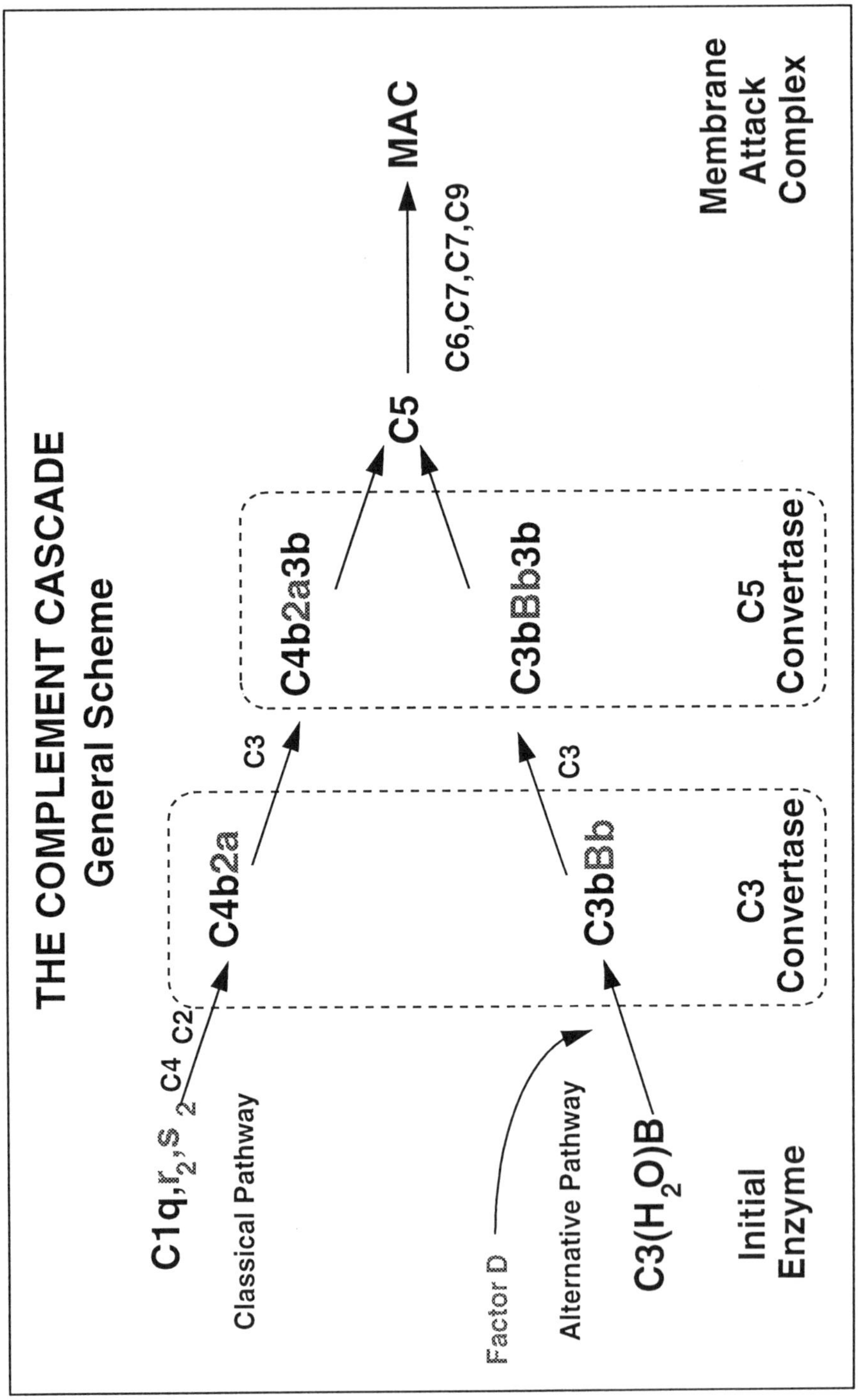

Fig. 10.1. Classical and alternative complement pathways. Adapted from Müller-Eberhardt HJ Ann Rev Biochem 1988; 57:321-347.

the C5 convertases in which the product of that reaction forms part of an enzyme complex acting upon C5. This is the key component of the final common pathway which generates the porin-like cytolytic membrane attack complex (MAC). It is not surprising, therefore, that regulatory molecules act at the level of both convertases and the final common pathway. In addition, complement activation generates chemically reactive protein intermediates (the thiol-ester containing species C3b and C4b which react covalently with nucleophiles in glycoproteins on cell membranes). It is equally unsurprising that regulators of complement activation are frequently cell membrane associated, however the definition of complement receptors needs clarification. As used here, it covers those molecules which interact at the cell surface with components of the activation pathways, regulate such activation and which link those pathways with other biological processes. This definition includes molecules which have no well-defined signaling function but important cofactor and negative regulatory actions. For simplicity, we have restricted discussion to those molecules generally termed complement receptors and have not included other important regulatory molecules such as decay accelerating factor (DAF, CD55) and membrane cofactor protein (MCP, CD46) which have mechanisms in common with CR1. We have also not considered CD59 which acts at the level of the membrane attack complex.

The regulation of complement activation (CA) needs to meet three key biological requirements:

1. Self/non-self discrimination.[5] The alternative pathway of CA is continuously activated by spontaneous hydrolysis of the thiolester bond in C3 and the generation of a C3(H20)-Factor B complex. This process must be restricted to non-self surfaces by the expression of self-encoded regulatory molecules.

2. Appropriate amplification. To achieve lethal cytolytic effects, CA pathways need amplification (a process readily mediated by linked protease cascades) but the generation of a rapid flux of MAC needs to be restricted in time as well as space to prevent destruction of nearby host cells.[6]

3. Non-depletion. As with blood coagulation, the protease cascades of CA have the potential to spiral out of control and consume key factors such as C3. Soluble regulatory factors (notably Factor H and C4bp) are required to prevent cell surface CA triggering general systemic complement activation. Although these can be recruited to cell surfaces, they are not 'receptors' in the sense used here.

4. Integration with cellular processes. The ability of CA to enhance immune responses in a ligand-specific manner and to activate and recruit neutrophils to sites of infection (and therefore CA) provides important synergies in the overall process of clearance of foreign substances.

10.2 COMPLEMENT RECEPTOR 1 (CR1/CD35)

10.2.1 STRUCTURAL ORGANIZATION

Complement receptor 1 is found in various polymorphic forms, the most common being the F (or A) allele which encodes a polypeptide of 190 kDa.[20] Analysis of the complete cDNA sequence[7,8] shows that the extracellular portion of the receptor is composed entirely of repeating protein units of approximately 60-70 amino acids, called short consensus repeats (SCRs). This type of domain is common to many complement proteins and non-complement related proteins[9] and is characterized by containing four invariant cysteines paired 1,3 and 2,4, as well as other highly conserved residues. The structure for SCRs 5, 15 and 16 of complement regulatory protein, Factor H, have now been solved and show a compact structure which is held together by a hydrophobic core and the disulphide bridges.[10,11]

The F allotype of complement receptor 1 contains 30 SCRs which can vary in identity from ~40% to 100%. Alignment of complement receptor 1 against itself shows a higher level of structural similarity; the sequence can be broken down into repeating units called long homologous repeats (LHRs) each composed of seven SCRs.[7] Hence the F allotype of CR1 has 4 LHRs followed by 2 SCRs, a transmembrane domain and a small cytoplasmic domain (Fig. 10.2).

Complement receptor type 1 inhibits complement activation by binding to the C3 and C5 convertases, promoting their dissociation and by acting as a cofactor for factor I protease which cleaves C3b and C4b. The high degree of identity between the SCR's coupled to the multiple functions carried out by CR1 has lead a number of groups to investigate the role of individual SCRs as well as the importance of specific amino acids within each SCR. Deletion mutagenesis of individual SCRs suggested that LHRA comprising SCR's 1-7 contained predominantly C4b binding activity and SCRs 1-2 played a key role in this binding. LHRB (SCRs 8-14) and LHRC (SCRs 15-21) were predominantly responsible for C3b binding.[12] SCRs 8-11 are virtually identical to SCRs 15-18 so it is not surprising that both LHRs possess a similar function. Overall the F allotype contains two C3b binding sites and one C4b binding site. A more detailed analysis showed that the first four SCR of LHRA contained all the information necessary to mediate C4b binding as well as cofactor I activity.[13-16] Likewise the first four SCRs of LHRB and LHRC contained all the information necessary to mediate C3b binding, some C4b binding activity as well as cofactor I activity.[13-17] Site-directed mutagenesis has shown there are specific amino acids within SCRs 1 and 2 that are required for C4b binding and similarly specific residues within SCRs 8-9 that mediate C3b binding. Substitution of these amino acids between SCRs can alter substrate specificity.[13,14] Hence, by key amino acid interactions and spreading binding across multiple SCRs, novel and specific

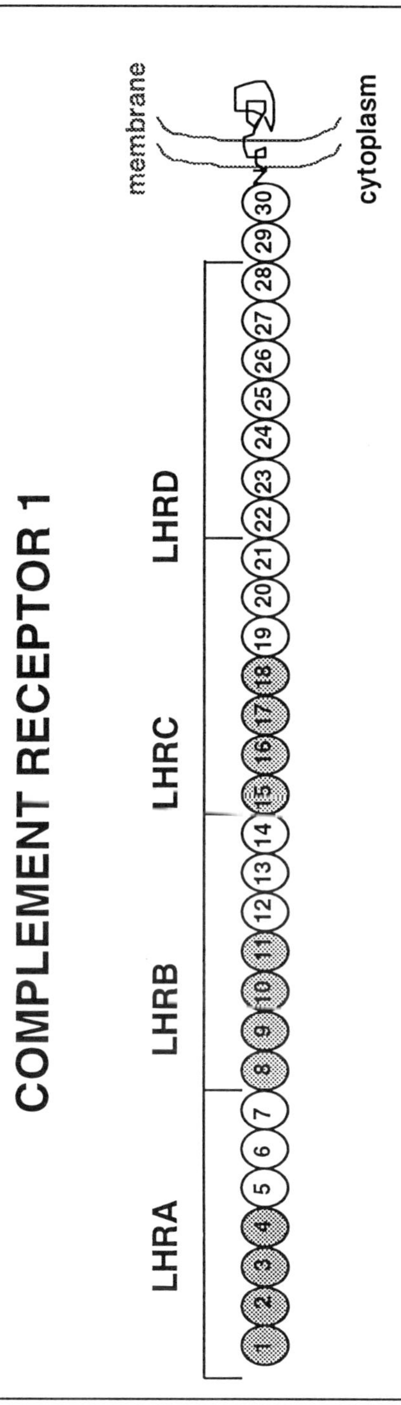

Alignments of the primary sequence show that the structure of CR1 can be divided into regions of homology called long homologous repeats (LHR). Each extracellular domain has a short consensus repeat type structure (SCR) which is ~ 60 - 70 amino acids in length.

Fig. 10.2.

interactions can me made by this class of structurally conserved protein module.

10.2.2 Chromosomal Localization of CR1 and Relevance to Function

Complement receptor 1 is located on chromosome 1 at position q32 within the RCA gene cluster (Regulation of Complement Activation) alongside other complement regulatory proteins such C4b binding protein, CR2, MCP, DAF and Factor H (reviewed in ref. 18). Each protein in this family is comprised of multiple repeats of the SCR type module, the smallest members, MCP and DAF, contain four SCR repeats. In general one exon encodes a single SCR although most of the characterized genes carry at least one exon split. The similarity of the coding sequences points to these multiple genes as having arisen by gene duplication from a single ancestor to modulate complement activation by a number of separate mechanisms. A more detailed analysis of the RCA gene cluster has shown that there are sequences related to CR1 and suggest that partial gene duplication has occurred.[19]

10.2.3 CR1 Polymorphisms

CR1 has several allelic forms A, B, C and D. The A (or F) form has an apparent M_R 190 kDa and a gene frequency of 0.83.[20] The B (or S) form, M_R 220 kDa, has a gene frequency of 0.16, the C form is 160 kDa and has a gene frequency of 0.01 and D form at 250 kDa, has a gene frequency of 0.001.[20] Speculation has been made that all these polymorphic variants, which differ by an internal repeat, have arisen as a result of unequal cross-over within repeated regions.[7,8,18] Deficiency in CR1 and or altered expression of CR1 have been implicated in a number of diseases such as systemic lupus erythematosus and rheumatoid arthritis[21] and genetic typing may be useful in disease predication.

10.3 COMPLEMENT RECEPTOR 2 (CR2/CD21)

Complement receptor 2 has a restricted distribution and is found on mature B lymphocytes, follicular and probably other dendritic cells and certain T-cell and epithelial cell subsets. It can form complexes with other cell surface proteins such as CD19 and CR1.[22,23] As isolated from the B lymphoblastoid Raji cell line, CR2 is a labile 140 kDa glycoprotein. CR2 possesses either 15 or 16 SCR repeats, the difference being due to alternative splicing[24] and appears to have a similar higher-order internal homology to that of CR1 (but based on 4-SCR rather than 7-SCR repeats). Like CR1, it has a single transmembrane domain and a short cytoplasmic domain. The latter contains tyrosine kinase and protein kinase C sites.[25] CR2 has two well-characterized ligands, the gp350/220 glycoprotein of Epstein-Barr virus (EBV)[25] and a proteolytic fragment of C3, C3d. There have also been reports that CR2 interacts with interferon-α[26] and CD23[27], the latter being a link to the control of IgE synthesis. CR2 appears to be the trophic factor for B cell infection by EBV leading to infectious mononucleosis and Burkitt's lymphoma. Binding of C3d is not associated with phagocytosis but rather appears to be linked to innate immunity. Workers in D.T. Fearon's laboratory have recently shown that recombinant constructs containing a model antigen and multiple copies of C3d appear to bind with high affinity to the Raji cell line and greatly stimulate production of antigen-specific IgG1 relative to antigen alone.[28] It was proposed that the covalent attachment of multiple copies of C3d (derived from Factor I proteolysis of deposited C3b) targeted the antigen to B cell precursors and directed the Ig response through the known interactions of CR2 and CD19.[23,29] Interaction with CR1 may also play a role because the C3d fragment is the product of a degradative process in which CR1 is key cofactor and by interacting with CR2, CR1 may 'hand on' a ligand. This mechanism provides a molecular basis for the recognition by the host of potentially dangerous non-self pathogens and also provides a clear link between the regulation of complement and cellular immunity.

10.4 COMPLEMENT RECEPTOR 3 (CR3)

Human complement receptor 3 (CR3, Mac-1, CD11b/CD18), is structurally unrelated to CR2 and CR1 and is not a member of the RCA gene cluster. It is expressed on phagocytic cells and lymphocytes and binds to the factor I generated fragment of C3b, iC3b, but not to C3b itself. It plays a major role in phagocytosis, cellular adhesion and has been demonstrated to be involved in complex receptor cross-talk and nuclear translocation events including those with urokinase receptor [CD87][30] and NF-κB[31] and Fc-γR.[32] CR3 is a bimolecular complex of two different gene products composed of a 95 kDa β-chain (CD18) which is common to other β-2 integrins, and a unique α-chain of 165 kDa (CD11b). LFA-1 is a related integrin whose α-chain is 185 kDa. Both CR3 and LFA-1 interact with the adhesion molecule ICAM-1 although CR3 interacts with the third immunoglobulin-like domain of ICAM-1 whereas LFA-1 interacts with the first immunoglobulin-like domain.[33] The key role of CR3 in phagocytosis is emphasized by many studies in which receptor deficiency has been linked to repeated bacterial infection.

A second protein p150,95 is now recognized as a functionally distinct complement receptor (CR4) although it clearly has a similar structure and ligand specificity to CR3. There is a recent report of a 16 kDa CR3-associated peptide which may be involved in the signaling process.[34]

Antibodies against CR3 or CR3 antagonists have been reported to have anti-inflammatory activity, for example in a rat colitis model[35] and the reverse passive arthus reaction.[36]

10.5 CR1 AS A THERAPEUTIC AGENT

The generation of a soluble form of CR1 was first reported by Weisman et al[37] who inserted a stop codon in the A (F) allotype of the CR1 gene immediately before the single transmembrane helix and generated a secreted protein comprising the 30 extracellular SCR domains (sCR1). This molecule was shown to retain the decay acceleration and Factor I cofactor activities of the parent CR1 and proved to be a potent inhibitor of complement activation in a variety of test systems in vitro.[37]

sCR1 (also known as TP10-HD and BRL55730) has been developed as a specific biopharmaceutical agent for the inhibition of inappropriate complement activation in man. A substantial number of animal pharmacology studies in various models of disease have demonstrated that the agent can suppress activation of both complement pathways when given systemically. The original observation[37] that suppression of the complement activation associated with reperfusion in a rat model of myocardial infarction resulted in reduced myocardial necrosis and infarct size has since been confirmed in other models of myocardial ischemia.[38] sCR1 has also been shown to be active in animal models of complement-mediated lung injury[39,40] and has demonstrated a protective effect against hyperacute vascular damage in several models of xenograft and allograft rejection.[41-45] sCR1 has come to be used as a convenient agent for probing the role of complement in the pathogenesis of disease. Unlike cobra venom factor (CVF) which turns over C3 through formation of a stable C3 convertase, sCR1 does not deplete the complement system of any components or generate anaphylatoxin species.

Recombinant human sCR1 has been produced at a scale and purity suitable for clinical studies and is currently in Phase II trials in acute myocardial infarction and adult respiratory distress syndrome.

As noted above, the repetitive domain structure of CR1 lends itself to a fragmentation approach aimed at assigning particular functions to specific SCRs or regions of sequence within SCRs. In our laboratory, we have studied a three-domain construct from the N-terminus of the molecule. We have shown that this protein can be expressed from a fully synthetic gene at high levels in bacteria and refolded from inclusion bodies in the reduced and denatured form to give an active complement inhibitor.[46] Surprisingly in view of earlier reports,[13,14] this fragment is active against both alternative and classical pathways and appears to retain Factor I cofactor activity, albeit at lower potency than sCR1 itself.

sCR1 and related molecules seem likely to form the first generation of complement inhibitor therapeutics suitable for human use. Further advances based on protein engineering of CR1 can be anticipated—particularly with respect to improved pharmacodynamics and targeting of inflammatory loci. Several chimeric molecules derived from other complement regulatory molecules such as CR1, DAF and CD59 have recently been described[47,48] but it is not yet clear what advantages will accrue from incorporation of more than one inhibitory mechanism into a single molecule.

Although CR1/CR2 chimeras have been described[17] use of soluble forms of CR2 as therapeutic agents has yet to be developed. In theory such agents could act as decoys for EBV or downregulate Ig responses through intervention in the C3d-CR2-CD19 cascade.

References

1. Morgan BP. Complement: Clinical Aspects and Relevance to Disease. London: Academic Press, 1990.
2. Sim RB, ed. Activators and Inhibitors of Complement. Dordrecht: Kluwer Academic Press, 1993.
3. Laws SKA, Reid KBM, Complement. Oxford: IRL Press, 1988
4. Whaley K, ed. Complement in Health and Disease. Lancaster: MTP Press, 1987.
5. Atkinson JP, Farries T. Separation of self from non-self in the complement system. Immunol Today 1987; 8:212-15.
6. Lublin D, Atkinson JP, Decay accelerating factor: biochemistry, molecular biology and function. Ann Rev Immunol 1989; 7:35-58.
7. Klickstein LB, Wong WW, Smith JA et al. Human C3b/C4b Receptor (CR1). Demonstration of long homologous repeating domains that are composed of the short consensus repeats characteristic of C3/C4 binding. J Exp Med 1987; 165:1095-1112.
8. Hourcade D, Miesner DR, Atkinson JP et al. Identification of an alternative polyadenylation site in the human C3b/C4b receptor (complement receptor type 1) transcriptional unit and prediction of a secreted form of complement receptor type 1. J Exp Med 1988; 168:1255-1270.
9. Reid BM, Day AJ. Structure-Function relationships of the complement components. Immunol Today 1989; 10:177-180
10. Norman DG, Barlow PN, Baron M et al. Three-dimensional structure of a complement control protein module in solution. J Mol Biol 1991; 219:717-725.
11. Barlow PN, Norman DG, Steinkasserer A et al. Solution structure of the fifth repeat of Factor H: A second example of the complement control protein module. Biochem 1992; 31:3626-3634.
12. Klickstein LB, Bartow TJ, Miletic VJ et al. Identification of distinct C3b and C4b recognition sites in the human C3b/C4b receptor (CR1/CD35) by deletion mutagenesis. J Exp Med 1988; 168:1699-1717.
13. Krych M, Hourcade D, Atkinson JP. Sites within the complement C3b/C4b receptor important for the specificity of ligand binding. PNAS USA 1991; 88:4353-4357.
14. Krych M, Clemenza L, Howdeshell D et al. Analysis of the functional domains of complement receptor type 1 (C3b/C4b receptor; CD35) by substitution mutagenesis. J Biol Chem 1994; 269:13273-13278.

15. Makrides SC, Scesney SM, Ford PJ et al. Cell surface expression of the C3b/C4b receptor (CR1) protects Chinese hamster ovary cells from lysis by human complement. J Biol Chem 1992; 267:24754-24761.
16. Reilly BD, Makrides SC, Ford PJ et al. Quantitative analysis of the C4b dimer binding to distinct sites on the C3b/C4b receptor (CR1). J Biol Chem 1994; 269:7696-7701.
17. Kalli KR, Hsu P, Bartow TJ et al. Mapping of the C3b-binding site of CR1 and Construction of a $(CR1)_2$-$F(ab)_2$ chimeric complement receptor. J Exp Med. 1991; 174:1451-1460.
18. Hourcade DM, Holers VM, and Atkinson JP. The regulators of complement activation (RCA) gene cluster. Adv Immunol 1989; 45:381- 416.
19. Hourcade D, Miesner DR, Bee C et al. Duplication and divergence of the amino-terminal coding region of the complement receptor 1 (CR1) gene. J Biol Chem 1990; 265:974-980.
20. Ross GD, Medof ME. Membrane complement receptors specific for bound fragments of C3. Adv Immunol 1985; 37:217-268.
21. Corvetta A, Pomponio G, Bencivenga R et al. Low number of complement C3b/C4b receptors (CR1) on erythrocytes from patients with essential mixed cryoglobulinemia, systemic lupus erythematosus and rheumatoid arthritis: relationship with disease activity, anticardiolipin antibodies, complement activation and therapy. J Rheumatol 1991; 18: 1021-1025..
22. Delibrias CC, Fischer E, Bismuth G et al. Expression, molecular association and functions of C3 complement receptors CR1 (CD35) and CR2 (CD21) on the human T cell line HPB-ALL. J Immunol 1992; 149:798-774
23. Fearon DT, Carter RH. The CD19/CR2/TAPA-1 complex on B lymphocytes: linking to natural immunity. Ann Rev Immunol 1995; 13:127-149
24. Fujisaku A, Harley JB, Frank MB et al. Genomic organisation and polymorphisms of the human C3d/Epstein-Barr virus receptor. J Biol Chem 1989; 264:2118-2125.
25. Ahearn JM, Fearon DT. Structure and function of the complement receptors CR1 (CD35) and CR2 (CD21). Adv Immunol 1989; 46:183-219.
26. Delcayre AX, Salas F, Mathur S et al. Epstein Barr virus/complement C3d receptor is an interferon alpha receptor. EMBO J 1991; 10:919-926
27. Aubry JP, Pochon S, Graber P et al. CD21 is a ligand for CD23 and regulates IgE production. Nature 1992; 358:505-507.
28. Dempsey PW, Allison ME, Akkaraju S et al. C3d of complement as a molecular adjuvant: bridging innate and acquired immunity. Science 1996; 271:348-350
29. Tuvenson DA, Ahearn JM, Matsumoto AK. Molecular interactions of complement receptors on B lymphocytes: A CR1/CR2 complex distinct from the CR2/CD19 complex. J Exp Med 1991; 173:1083-1089.
30. Gyetko MR, Sitrin RG, Fuller JA et al. Function of the urokinase receptor (CD87) in neutrophil chemotaxis. Leukoc Biol 1995; 58:533-538.
31. Thieblemont N, Haeffner-Cavaillon N, Haeffner et al. Triggering of

complement receptors CR1 (CD35) and CR3 (CD11b/CD18) induces nuclear translocation of NF-kappa B (p50/p65) in human monocytes and enhances viral replication of HIV infected monocytes. J Immunol 1995; 155:4861-4867.

32. Ohkuro M, Ogur-Maski M. Kobayashi K et al. Effect of iC3b binding to immune complexes upon the phagocytic response of human neutrophils: synergistic functions between Fc gamma R and CR3. FEBS Lett 1995; 373:189-192.
33. Diamond MS, Staunton DE, Marlin SD et al. Binding of the integrin Mac-1 (CD11b/CD18) to the third immunoglobulin-like domain of ICAM-1 (CD54) and its regulation by glycosylation. Cell 1991; 65:961-971.
34. Messika EJ, Avni O, Gallily R et al. Identification and characterisation of a novel protein associated with macrophage complement receptor 3. J Immunol 1995; 154:6563-6570.
35. Palmen MJ, Dijkstra CD, Ende MB et al. Anti-Cd11b/CD18 antibodies reduce inflammation in acute colitis in rats. Clin Exp Immunol 1995; 101:351-356.
36. Rote WE, Dempsey E, Maki S et al. The role of CD11/CD18 integrins in the reverse passive arthus reaction in rat dermal tissue. J Leukocyte Biol 1996; 59:254-261.
37. Weisman HF, Bartow L, Leppo MK et al. Soluble human complement receptor type 1 : in vivo inhibitor of complement suppressing post-ischemic myocardial inflammation and necrosis. Science 1990; 249: 146-51
38. Smith EF 3d, Griswold DE, Egan JW et al. Reduction of myocardial infarction with human soluble complement receptor 1 (BRL 55730). Eur J Pharmacol 1993; 236:477-81
39. Rabinovici R, Sofronskiv ND, Borboroglu P et al. Interleukin-2 induced lung injury. The role of complement. Circulation Research 1994; 74:329-35
40. Mulligan MS, Yeh CG, Rudolph AR et al. Protective effects of soluble CR1 in complement and neutrophil mediated tissue injury. J. Immunol 1992; 148:1479-1485.
41. Pruitt SK, Bollinger RR. The effect of soluble complement receptor 1 on hyperacute allograft rejection. J Surg Res 1991; 50:350-355
42. Pruitt SK, Baldwin WM. The effect of soluble complement receptor type 1 on hyperacute xenograft rejection. Transplantation 1991; 52:868-873
43. Xia W, Fearon DT, Moore FD Jr et al. Prolongation of guinea pig cardiac xenograft survival in rats by soluble human complement receptor type 1. Transplant Proc 1992; 24:479-480.
44. Pruitt SK, Kirk AD, Bollinger RR et al. The effect of soluble complement receptor 1 on hyperacute rejection of porcine xenografts. Transplantation 1994; 57:363-370.
45. Pruitt SK, Bollinger RR, Collins et al. Continuous complement inhibition using soluble complement receptor 1 (sCR1); effect on hyperacute rejection (HAR) of pig to primate cardiac xenografts. Transplant Proc 1996; 28:756.

46. Dodd I, Mossakowska DE, Camilleri P et al. Over expression in Escherichia coli, Folding, purification and characterisation of the first three short concensus repeat modules of human complement receptor 1. Protein Expression and Purification 1995; 6:727-736
47. Fodor WL, Rollins SA, Guilmette ER et al. A novel bifunctional chimeric complement inhibitor that regulates C3 convertase and formation of the membrane attack complex. J Immunol 1995; 155:4135-4138.
48. El Feki SG, Fearon DT. CR1-CD59 chimera: a novel recombinant regulator of complement activation. Mol Immunol supp 1996; 33:57.

INDEX

E

F

G

H

I

J

K

L

M

N

O

P

Q

R

S

T

U

V

X

Z